T0214524

Lecture Notes
in Business Information Processing   345

More information about this series at http://www.springer.com/series/7911

Nikolay Mehandjiev · Brahim Saadouni (Eds.)

Enterprise Applications, Markets and Services in the Finance Industry

9th International Workshop, FinanceCom 2018
Manchester, UK, June 22, 2018
Revised Papers

 Springer

Editors
Nikolay Mehandjiev
Alliance Manchester Business School
Manchester, UK

Brahim Saadouni
Alliance Manchester Business School
Manchester, UK

ISSN 1865-1348   ISSN 1865-1356   (electronic)
Lecture Notes in Business Information Processing
ISBN 978-3-030-19036-1   ISBN 978-3-030-19037-8   (eBook)
https://doi.org/10.1007/978-3-030-19037-8

This Springer imprint is published by the registered company Springer Nature Switzerland AG
The registered company address is: Gewerbestrasse 11, 6330 Cham, Switzerland

Preface

Advancements in information and communication technologies have paved the way to new business models, markets, networks, services, and players in the financial services industry, developments now labelled as "FinTech." FinanceCom workshops have been providing research leadership in this area well before the rise of the FinTech concept, helping academics and practitioners to understand, drive, and exploit the associated opportunities in the financial sector.

After very successful FinanceCom workshops in Sydney (twice), Regensburg, Montreal, Paris, Frankfurt (twice), and Barcelona, FinanceCom 2018 was held in Manchester (UK) for the first time. The workshop spans multiple disciplines related to the use of technology for financial services, including analytical, technical, services, economic, sociological and behavioral sciences, aiming to foster true cross-disciplinary work for addressing research questions grounded in real industrial needs. Indeed, one of the main goals of the workshop series has been to bridge the boundaries between "traditional" business disciplines (such as finance and economics) and "ICT-based" disciplines (such as software engineering and information systems), which demonstrate inertia in their focus on isolated parts of a much larger picture in which diverse aspects are intertwined in complex ways.

This inertia reduces the size of the FinanceCom community yet at the same time it amplifies the impact of the work presented at our specialized interdisciplinary workshop series. This year we received 18 submissions, of which we selected 11 high-quality papers to be presented at the workshop and to consequently be reworked based on the feedback and published in this volume together with an additional invited paper. The selection was instrumented by a rigorous review process implemented with the help of a Program Committee consisting of internationally renowned researchers and practitioners.

The first part of these proceedings contains five contributions in the area of financial innovation. The first paper is our invited paper "The MiFIR Trading Obligation: Impact on Trading Volume and Liquidity in Electronic Trading" by Gomber et al. It tests whether the migration of trading from the over-the-counter (OTC) markets to regulated trading venues and on platforms of systematic internalizers (Sis) has any effect on liquidity. The paper shows that liquidity on open limit-order book markets (OLOBM) increases as a result of the additional turnover coming from the OTC market. However, trading on SIs has an adverse effect on liquidity for OLOBM. The second paper, "Open Innovation Effectiveness in the Financial Services Sector," by Piobbici et al. argues that an open innovation framework could be useful in analyzing innovation patterns that involve both financial and non-financial institutions. The paper suggests that market-based collaborations are significantly correlated with the performance in relation to innovation. Overall, the paper demonstrates that market-based collaborations are the most effective. Using a very large sample of 2469 observations and covering a 13-year period, the third paper, "Identification of Financial Statement Fraud in Greece by Using Computational Intelligence Techniques," by Chimonaki et al. reports that computational intelligence (CI) techniques can enhance the detection of financial fraud. The fourth paper, "What Sort of Asset? Bitcoin Analyzed," by Corbet et al. examines whether the introduction of futures contracts in bitcoin addresses the issues that have inhibited it from being classified as a currency. The paper shows that spot volatility has increased following the introduction of futures contracts, these contracts are ineffective as hedging instruments, and price discovery is driven by uninformed investors in the spot market. Overall, the paper demonstrates bitcoin is a speculative asset and the introduction of futures contracts has failed to make it a currency. The fifth paper, "Blockchained Sukuk-Financing," by Shaikh and Zaka examines the use of *sukuk* in Islamic finance as an alternative source of finance to conventional bonds. This source of finance can involve a number of parties, especially when sale, lease, and agency contract are combined. The paper presents a model for blockchained *sukuk*-issue, highlighting the significant design features that are unique for this source of finance. This approach should help the traceability of any asset transfers that will ultimately enhance *sukuk* credibility and their valuation. Further, a better contract infrastructure with blockchain security should also significantly reduce the execution time for transactions involving *sukuk*.

The second part of the proceedings contains four contributions in the area of market data analytics. The first paper in this section, "The Role of Customer Retention in Business Outcomes of Online Service Providers," by Assemi examines two competing theoretical models using archival data from a leading crowdsourcing marketplace. The findings show that a provider's profile information is significantly associated with the provider's business outcomes, while customer retention partially mediates this relationship. Furthermore, the results demonstrate the important role of new customers in achieving better business outcomes on crowdsourcing marketplaces. The second paper, "Using NSIA Framework to Evaluate Impact of Sentiment Datasets on Intraday Financial Market Measures: A Case Study," by Qudah and Rabhi examines the impact of news sentiment on the intraday cumulative average abnormal returns (CAAR) using cases from Australia, Canada, Germany, and the USA. The results show that in nine out of the 12 cases examined, the release of the sentiment news has a significant negative impact on the CAAR. The third paper, "Financial Data Visualization in 3D on Immersive Virtual Reality Displays," by Lugmayr et al. presents a design prototype for the 3D visualization of financial data relating to Australia's energy sector with a large-scale immersive virtual reality environment. The paper attempts to develop a proof-of-concept implementation referred to as "ElectrAus," a tool to visualize publicly available data from the Australian energy market. The primary aim of the proof-of concept implementation is to demonstrate new concepts in data visualization such as utilization of immersive environment and large screens for financial data visualization; creating an understandable visualization for the general public; and additional insights and analytics through 3D displays. The fourth paper, "Document Representation for Text Mining: Opportunities for Analytics in Finance," by Roeder and Palmer investigates the utilization of a distributed representation of words and documents referred to as "embeddings" for text analytics in finance. The results reveal a potential application of the document representation techniques for text analytics in the area of finance.

The third part of the proceedings contains three papers focusing on the use of semantic modelling in supporting financial trading. In their paper, "Semantic Model Based Framework for Regulatory Reporting Process Management," Pilaka et al. describe how semantic modelling can help in extracting instances of regulatory reporting processes from event traces and help with compliance monitoring. The proposed framework is tested through applying it to event traces from the Australian Securities Exchange to extract instances of the "off market bid" regulatory process. The second paper, "Applying Ontology-Informed Lattice Reduction Using the Discrimination Power Index to Financial Domain," by Quboa et al. describes the use of semantically encoded knowledge about asset allocation to support the automatic clustering of instances sampled from the domain and their tagging with semantic information. The final paper in this section by Behnaz et al., "A Statistical Learning Ontology for Managing Analytics Knowledge," proposes an ontology development process tuned to developing statistical learning ontologies that can support analytics. Two case studies ground the research to the domains of commodity pricing and digital marketing.

Special thanks go to Alliance Family Foundation and Alliance Manchester Business School Strategic Investment Fund, which supported this event financially through the Alliance MBS Big Data Forum. We are also grateful to Fethi Rabhi, who has guided us all the way in getting this workshop and proceedings organized from start to finish. We are grateful to our team of reviewers and Program Committee members, who worked very hard with the authors to ensure the quality of the papers included in this volume, and to Ralf Gerstner and Christine Reiss from Springer for their excellent support in producing this proceedings volume.

June 2018

Nikolay Mehandjiev
Brahim Saadouni

Organization

The workshop took place at the Alliance Manchester Business School in Manchester, UK. Financial support by the Alliance Family Foundation through the Alliance MBS Big Data Forum is gratefully acknowledged.

Organizing Committee and Program Chairs

Nikolay Mehandjiev	Alliance Manchester Business School, UK
Brahim Saadouni	Alliance Manchester Business School, UK

Program Committee

Marc Adam	University of Newcastle, Australia
Madhushi Bandara	University of New South Wales, Australia
Sonia Cisneros-Cabrera	Alliance Manchester Business School, UK
Onur Demirors	Izmir Institute of Technology, Turkey
Saif Dewan	Australian National University, Australia
Stefan Feuerriegel	University of Freiburg, Germany
Mahdi Fahmideh Gholami	University of Wollongong, Australia
Peter Gomber	University of Frankfurt, Germany
Nikolay Kazantsev	Alliance Manchester Business School, UK
Stefan Lessmann	Humboldt University of Berlin, Germany
Artur Lugmayr	Curtin University, Australia
Jan Muntermann	University of Göttingen, Germany
Dirk Neumann	University of Freiburg, Germany
Maurice Peat	University of Sydney, Australia
Helmut Prendinger	National Institute of Informatics, Japan
Qudamah Quboa	University of Manchester, UK
Fethi Rabhi	University of New South Wales, Australia
Federico Rajola	Catholic University of the Sacred Heart, Italy
Ryan Riordan	University of Ontario, Canada
Michael Siering	University of Frankfurt, Germany
Andrea Signori	Università Cattolica del Sacro Cuore, Italy
Basem Suleiman	University of Sydney, Australia
Christof Weinhardt	Karlsruhe Institute of Technology, Germany
Axel Winkelmann	University of Würzburg, Germany

Steering Committee for the FinanceCom Workshop Series

Christof Weinhardt	Karlsruhe Institute of Technology, Germany
Dennis Kundisch	University of Paderborn, Germany

Federico Rajola   Catholic University of Milan, Italy
Fethi Rabhi   University of New South Wales, Australia
Jan Muntermann   University of Frankfurt, Germany
Peter Gomber   University of Frankfurt, Germany
Ryan Riordan   Karlsruhe Institute of Technology, Germany

Contents

Semantic Modelling

Financial Innovation

The MiFIR Trading Obligation: Impact on Trading Volume and Liquidity in Electronic Trading

Peter Gomber[ID], Benjamin Clapham[ID], Jens Lausen[ID], and Sven Panz[(✉)][ID]

Goethe University Frankfurt, Theodor-W.-Adorno-Platz 4,
60323 Frankfurt am Main, Germany
{gomber,clapham,lausen,panz}@wiwi.uni-frankfurt.de
http://www.efinance.wiwi.uni-frankfurt.de

Abstract. The new financial market regulation MiFID II/MiFIR will fundamentally change the trading and market infrastructure landscape in Europe. One key aspect is the trading obligation for shares that intends to restrict over-the-counter (OTC) trading to ensure that more trading takes place on regulated trading venues and on platforms of Systematic Internalisers (SIs). In this context, market experts often argue that SIs might have a competitive advantage due to the best execution concept in combination with the possible exemption of SIs from the tick size regime. Applying scenario analysis, we determine the likely migration of OTC trading volume to regulated trading venues and SIs. Based on our data set, we investigate how changes in trading volume influence liquidity on open limit order book markets (lit markets). The results of our scenario analysis indicate that liquidity on lit markets might increase due to additional turnover formerly traded OTC. However, also a negative liquidity effect for lit markets and for the price discovery process is possible because of increased trading via SIs.

Keywords: MiFID II/MiFIR · Trading obligation · Liquidity ·
Trading volume · Electronic trading

1 Introduction

Electronic trading in European equity markets has changed significantly in the last 15 years due to regulatory initiatives. The Markets in Financial Instruments Directive (MiFID I) came into effect in April 2004 and has been applied by investment firms and Regulated Markets (RMs) in Europe since November 2007. A central goal of MiFID I was to enable investors to "trade securities at maximum efficiency and at minimum cost" [10]. This goal should be achieved by increased transparency and accessibility of markets, investor protection, market

---

The authors acknowledge financial support from Deutsche Boerse AG.

N. Mehandjiev and B. Saadouni (Eds.): FinanceCom 2018, LNBIP 345, pp. 3–26, 2019.
https://doi.org/10.1007/978-3-030-19037-8_1

integrity, harmonized European regulation and a level playing field among different types of trading venues to assure competition and to foster innovation. Furthermore, MiFID I increased competition between trading venues through the introduction of multilateral trading facilities (MTFs).

The competition between trading venues fostered by MiFID I led to a highly fragmented equity market in Europe. As of June 2017, new competitors of the incumbent national exchanges together achieved a market share of more than 28% of total European electronic order book trading in equities [17]. Industry studies (e.g., [32]) show that the service and fee competition triggered by MiFID I reduced explicit transaction costs both at trading venues and post-trade infrastructures.

Today, new requirements laid down in MiFID II and its accompanying regulation Markets in Financial Instruments Regulation (MiFIR) seek to extend the benefits that MiFID I generated for equity markets to other asset classes and to address problems caused by market fragmentation, dark trading, and over-the-counter (OTC) trading. One of the key objectives of MiFID II and MiFIR, which have to be applied from January 3rd, 2018, is that OTC trades (unless they fulfill certain criteria as discussed in detail in Sect. 3) are being forced to take place on RMs, MTFs, or Systematic Internalisers (SIs) due to the so-called trading obligation for shares. The goal of this study is to describe the context of the new trading obligation for shares and to investigate its potential effect on trading volume and liquidity on *lit venues*, i.e. transparent open order book markets that play a crucial role in the price discovery process. Thereby, this study should serve as a toolbox for market participants to assess the impact of the new regulation based on their own individual estimates regarding likely scenarios and the thresholds for migrating OTC volumes. Based on data and experts' estimates collected by a questionnaire, this study provides a prospective assessment of potential effects due to the introduction of the MiFIR trading obligation.

In Sect. 2, we discuss the drivers and goals of MiFID II/MiFIR against the background of changes in European equity trading due to MiFID I, technological developments in trading in the last ten years, and lessons learned from the financial crisis. Section 3 describes the concept of the MiFIR trading obligation and the related industry discussions concerning its implementation in detail. In Sect. 4, we explain the scenario methodology and develop the scenarios to assess the potential impact of the trading obligation on market share distribution and market liquidity in European equity trading. Section 5 describes our empirical analysis and results regarding the impact of the trading obligation on trading volume and liquidity. Section 6 discusses the results as well as limitations of this study. Finally, Sect. 7 concludes this paper.

2   Analysis of Drivers and Goals of MiFID II/MiFIR

Against the background of the financial crisis, new trading technologies, new products, and the necessary reviews already pre-defined in Article 65 of MiFID I

[14], the European Commission triggered a massive overhaul of European securities legislation. This initiative resulted in the publication of MiFID II and the accompanying regulation MiFIR in June, 2014 [15,16]. Starting on January 3rd, 2018, the new rules had to be applied by investment firms, market operators, data reporting service providers, and specific third-country firms. MiFID II and MiFIR aim at (i) assuring a safer, sounder, more transparent, and more responsible financial system, (ii) contributing to the delivery of the G20 commitment to tackle less regulated and opaque parts of the financial system, (iii) improving the organization and transparency of markets especially in those instruments traded mostly OTC, (iv) improving the oversight and transparency of commodity derivative markets, (v) taking account of new developments in market structures and technology, like dark trading, algorithmic trading, and high frequency trading (HFT), and (vi) minimizing the discretions available to member states [11].

The regulatory changes introduced by MiFID II/MiFIR also reflect the results of various academic studies that analyze the impact of MiFID I on market quality. A comprehensive overview of these studies is provided by [22]. While the majority of studies finds that MiFID I indeed broke national monopolies of exchanges thereby lowering transaction costs and also spurring new technical developments, most studies find no positive effect of MiFID I on transparency and integrity of European financial markets [22].

Although MiFID I changed the competitive landscape in lit trading considerably, OTC trading still represents a high and very relevant market share in total European equities trading contradicting one of the key objectives of MiFID I, which is to increase trading transparency. Analyzing a large intraday data set of OTC transactions from January 2008 to June 2013 covering EURO STOXX 50 constituents, [23] finds that OTC trading accounts for more than 50% of total trading volume with increasing shares during the sample period. Furthermore, their results reveal that the majority of OTC trades (58%) are retail-sized trades while less than 5% are above the large-in-scale (LIS) threshold according to the definition of MiFID I. The fraction of LIS-trades has decreased from more than 15% in 2008 to even less than 2% in 2013. In 2013, more than 95% of OTC trades would not have had a market impact if they had been submitted to the main market. Contrary to the MiFID I spirit and text, these results show that OTC trades are not infrequent and large, but are rather frequent and small. In addition, the increased emergence of dark pools, especially OTC dark pools or so-called broker crossing networks (BCNs)[1], harm liquidity since they do not disclose pre-trade information [9].

To achieve the goals and to address the weaknesses of MiFID I, several adjustments and extensions have been implemented within MiFID II/MiFIR. In particular, transparency goals were not completely achieved with MiFID I since trading systems such as BCNs emerged, which were not captured by the

---

[1] CESR, the predecessor of ESMA, defined Broker Crossing Networks as "internal electronic matching systems operated by an investment firm that execute client orders against other client orders or house account orders" [4].

regulatory regime, and a large fraction of trading is still conducted OTC. Consequently, MiFID II and MiFIR include several new regulations aiming at enhancing transparency, which are described in the following Sect. 3.

3   MiFIR Trading Obligation and Related Regulatory and Industry Debate

A key concept of MiFID II/MiFIR is to require all organized trading to take place on organized venues and to ensure that trading systems are properly regulated to assure a higher level of transparency (MiFIR, Recital 6). To achieve this goal, the new market framework introduces four key concepts: (i) a new category of trading venues called Organised Trading Facility (OTF) for non-equity instruments to be traded on a multilateral platform, (ii) the so-called "double volume cap regime" for equity trading, (iii) a new trading obligation for derivatives, and (iv) a new trading obligation for shares. Moreover, transparency requirements that only apply to shares according to MiFID I are extended to equity-like and non-equity instruments establishing uniform requirements for the transparency of transactions in financial markets (MiFIR, Recital 1).

The trading obligation for shares intends to restrict equity trading conducted OTC to ensure that more trading takes place on regulated trading venues and on platforms of SIs in order to increase transparency and to improve the quality of the price discovery process as well as liquidity on lit markets (MiFIR, Recital 11). It requires investment firms to undertake all trades in shares on an RM, an MTF, an SI, or an equivalent third-country trading venue. Exemptions to the trading obligation only apply if there are legitimate reasons[2]. MiFIR, Recital 11 explicitly states in this context that "[...] an exclusion from that trading obligation should not be used to circumvent the restrictions introduced on the use of the reference price waiver and the negotiated price waiver or to operate a broker crossing network or other crossing system". Therefore, BCNs, that are a relevant part of OTC trading in equities today, cannot exist any longer in their former set-up after January 3rd, 2018. These systems have to be authorized and operated as MTFs or SIs (MiFIR, Recital 6).

The introduction of the trading obligation and the prohibition of BCN triggered an intensive industry and regulatory debate concerning the level playing field and delineation between bilateral (SI and OTC) and multilateral trading (RM and MTF). This debate centered around the possibility for SIs to transact via riskless back-to-back transactions both by executing customer orders

---

[2] According to MiFIR Article 23 (1), trades can only be executed on an OTC basis if they are non-systematic, ad-hoc, irregular and infrequent, or are carried out between eligible and/or professional counterparties and do not contribute to the price discovery process. Article 2 of Commission Delegated Regulation (EU) 2017/587 (RTS 1) lists seven circumstances where trades do not contribute to the price discovery process: vwap-twap trades, portfolio trades, hedges, transfers among fund portfolios, give-ups/give-ins, collateral transfers, deliveries in case of exercises, securities financing transactions and buy-ins.

within their own SI and with other SIs or HFT Market Makers via so-called SI-Networks. Exchanges and various regulators argued that this might create unfair advantages for SIs relative to multilateral venues, reduce market transparency and impede the efficiency of the price discovery process. After an intensive correspondence between ESMA, the European Parliament and the European Commission in early 2017, the European Commission proposed a Commission Delegated Regulation [12] to clarify and amend the definition of SIs by adding a new article 16(a) to Commission Delegated Regulation No 2017/565 that states: *"An investment firm shall not be considered to be dealing on own account [...] where that investment firm participates in matching arrangements entered into with entities outside its own group with the objective or consequence of carrying out de facto riskless back-to-back transactions in a financial instrument outside a trading venue"*. Recital 2 of the Commission Delegated Regulation applies this restriction to both internal and external matching of trades, thereby referring to the intensively discussed networks of SIs.

This debate reveals that MiFID II/MIFIR re-launch a fight for market shares between trading venues and SI/OTC trading on the one hand and among the different types of trading venues on the other hand. Therefore, it is by no means a given fact that the trading obligation and the prohibition of BCNs will result in higher market shares of multilateral venues relative to bilateral means of execution.

An intelligent interpretation and implementation of the new MiFID II/MIFIR set-up by the financial industry might even result in higher shares of non-lit and non-multilateral executions in European equity trading from January 2018 onwards. Specifically, the tick size regime in combination with best execution requirements and the double volume caps within MiFID II/MiFIR might lead to a shift of trading behavior and resulting order flows.

In order to curb competition among trading venues based on tick size, MiFID II (Article 49) imposes the adoption of a (minimum) tick size regime that is defined in prescriptive and detailed tick size tables of Commission Delegated Regulation No 2017/588 [13]. While Article 49 MiFID II requires RMs to comply with the tick size regime, Article 18 (5) extends this requirement to MTFs (and OTFs). However, the tick size regime does not extend to SIs and OTC trading. Empirical research of [24] reveals that 29.3% of all SI trades and 24.8% of all OTC trades in a 2.5 years data set of European top liquid stocks violate the existing voluntary tick size agreements [18] among exchanges and MTFs in Europe. The vast majority of those tick size violations are trivial price improvements over the quotes of public lit markets. The ability to grant those trivial price improvements in the SI space and thereby still meeting the requirements of the trading obligation makes the SI regime very attractive. As it enables to execute within the public spread and to free-ride on the public price discovery process, it reduces the incentives of liquidity providers to provide competitive quotes in lit markets. Furthermore, the price improvements, even if economically very small, will likely prioritize SIs over lit markets in order routing decisions driven by best execution requirements. In response to ongoing debates about the minimum tick size exemption for SIs, ESMA opened a public consultation in

November 2017 to clarify that SIs' quotes should reflect the tick sizes applicable to EU trading venues.

The dark pool caps will also result in a change of trading behavior. Market operators (e.g., Cboe Europe[3] offering a Periodic Auctions Book model) as well as buy and sell side firms (e.g., the Plato Partnership project that is linked to the Turquoise Block Discovery service) launch new services and functionalities that help to avoid a classification of orders under the negotiated trade and reference price waivers and classify them as executions under either the auction model (and therefore limited pre-trade transparency) or under the LIS waiver (that is not included in the double volume cap mechanism). While this will increase the relevance of block trading in the future European execution ecosystem, investors will still try to identify ways to execute below LIS trades without pre-trade transparency and market impact. Against the background of the BCN prohibition, SIs offer an attractive alternative for customers in this respect as the SI transparency is limited to standard-market sizes (which is 7,500 € for most European shares), SIs fulfill the trading obligation requirements, and SIs enable customers to receive price improvements in combination with lower explicit fees compared to trading venues. The concept enables SI providers to keep order flows currently traded in their BCNs while at the same time fulfilling the requirements of the trading obligation (without the need to become an MTF). Moreover, they are able to tailor quotes to specific clients based on risk considerations, to cream-skim the order flow, i.e. to send informed order flow to public markets, and to attract smart order routers based on small increments in price improvements over the public spread. SI executions (e.g., at midpoint) also do not count under the reference price waiver and are not included in the double volume caps. The benefits of the SI regime might additionally attract current liquidity providers in BCNs to register as SIs in an attempt to retain their business model.

Thereby, the new regulatory framework and the combined effect of the trading obligation, the tick size regime, and the double volume cap will significantly influence European equities trading and will likely result in a redistribution of market shares between multilateral trading venues, SI, and OTC trading.

4   Scenario Development

The scenario analysis is a qualitative forecasting technique that is useful for strategic planning [3]. It is an effective methodology to deal with uncertainty when forecasting complex developments or structures, which is regularly used by academics and practitioners [34]. The scenario analysis represents an alternative to extrapolating past trends and relationships. In line with [29], we understand scenarios as "a narrative description of a consistent set of factors which define in a probabilistic sense alternative sets of future business conditions". A scenario analysis enables to capture a whole range of possible future outcomes thereby revealing how different assumptions and uncertainties interact under certain conditions [36]. Scenarios are suitable to capture new states after major

---

[3] Formerly Bats Europe.

shocks, e.g., due to new regulation or technological changes. When conducting a scenario analysis, it is essential to clearly define the scope, to identify trends as well as key uncertainties, and to consider major stakeholders, who will be affected by possible changes [36]. To derive realistic scenarios regarding the impact of the MiFIR trading obligation on market share distribution and liquidity, we rely on the one hand on historical trends and relationships and on the other hand, we make additional assumptions to cover future developments.

We prospectively evaluate three different scenarios to assess potential migrations of trading volume and related effects on liquidity due to the trading obligation imposed by MiFIR Article 23. These three scenarios reflect different reactions of relevant stakeholders of the financial industry to cope with the new regulation from January 2018 onwards. This allows to assess the effect of the trading obligation also with varying expectations in this regard. The three scenarios of our study are defined as follows:

- Scenario A (*"Pro-Multilateral"*) assumes that OTC trades (including the BCN share of OTC trading) will migrate to lit markets according to different trade size categories and their probability of migration.
- Scenario B (*"BCN Volumes Migrate to SIs"*) assumes that the providers of BCNs will manage to migrate the current BCN trading volumes into new or existing SI setups. Therefore, in scenario B, we assume that OTC trades (excluding the BCN share of OTC trading) will migrate to lit markets according to different trade size categories and their probability of migration.
- Scenario C (*"Internalization as in the US"*) assumes on the one hand that new realizations and implementations of the SI concept will enable providers of BCNs and other SIs to migrate the current BCN trading volumes into new or existing SI setups. On the other hand, additional volume from lit markets is expected to migrate into those SI setups, e.g., due to the possible ability to price improve relative to the tick size regime and due to related best execution obligations. Therefore, in scenario C, we assume that (i) OTC trades (excluding the BCN share of OTC trading) will migrate to lit markets according to different trade size categories and their probability of migration and (ii) new SI set-ups will attract trading volume equivalent to the US BCN share plus the US retail internalization level adjusted to the European context.

In order to assess the likelihood of migration for different OTC trade size categories, we conducted an online survey among industry experts on trading and market structure. This survey was composed of 25 questions dealing with different aspects of MiFID II/MiFIR that might have an impact on European market structure. Access to the questionnaire started on November 29th, 2017 and closed on December 19th, 2017. This ensures that participants' answers are based on a prospective view and are not biased through first changes triggered by the new rules. The questionnaire was distributed among 375 experts, whereof 111 experts submitted the online questionnaire before the deadline. This results in a response rate of 29.6%. Among more general questions, the survey participants were specifically asked to evaluate the scenarios described above and to estimate the volume shifts of OTC volumes in different trade size categories to lit markets due to the trading obligation for shares.

We assume that the probability of OTC volume migrating to lit markets depends on the respective size of an OTC trade and 98% of the respondents agreed to this assumption. The survey participants estimated that on average 39% of the OTC trades smaller than 500,000 € and 22% of the OTC trades between 500,000 € and 50 mn € will migrate to lit markets. As OTC trades larger than 50 mn € are very likely to account for portfolio transfers and other non-trading related motivations, a vast majority of survey participants (98%) expects that this kind of trades will remain OTC. Therefore, we assume that no volume of this category will shift to lit markets.

Additionally, the participants were asked to determine which of the three scenarios (A-C) appear to be most realistic. 17% of the survey participants consider scenario A to be most realistic while 34% of the respondents believe that scenario B is most realistic. With 43%, the largest group of industry experts considers scenario C to be most realistic among the three different scenarios. Only 5% of the respondents think that none of the scenarios described above is realistic[4].

5 Empirical Analysis

5.1 Data Set

In the next steps, we determine the likely change in trading volume on lit markets due to the trading obligation for OTC trades respectively the BCN trading prohibition for the three scenarios described above. Furthermore, we investigate the liquidity effect due to these volume shifts in public limit order books on the main market. For this analysis, we use Thomson Reuters Tick History (TRTH) as the primary source of data. Our sample comprises constituents of the EURO STOXX 50 as of January 2013. The sample period covers all trading days between 1 January 2008 and 30 June 2013. The sample period is suitable for the research question at hand since it ends one year before MiFID II came into force. Thus, it excludes any possible announcement effects or changes in trading protocols to cope with the new regulatory environment such as new mechanisms to handle the double-volume cap mechanism that were already introduced shortly after MiFID II came into force in mid-2014 (e.g., the Volume Discovery Order, Cboe Europe's Periodic Auctions Book, or the Plato Partnership). Nevertheless, market share distribution and overall trading volume in the last quarter of our sample (2nd Quarter 2013), which serves as the reference for our scenario analysis, were similar to the same quarter in 2017 (2nd Quarter 2017) before the application of MiFID II/MiFIR.

Our coverage of venues includes the main market for each stock of the EURO STOXX 50 index. Therefore, we include data of the following main markets:

---

[4] The complete analysis of the survey results can be found in [21].

Paris, Frankfurt, Amsterdam, Milan, Helsinki, Madrid, Brussels and Dublin. The data contains trades on all these main markets as well as order book updates at a millisecond precision. In addition to the main markets, we include all trades from alternative venues such as Bats, Chi-X[5] (now Cboe BXE and Cboe CXE) and Turquoise, thus covering more than 95% of overall trading volume of these stocks on lit markets. Besides the trades executed in open limit order books, we include all off-exchange trades reported to Markit's trade reporting platform BOAT.

5.2 Volume Migration According to the Scenarios

Our empirical analysis comprises two consecutive steps. In step one, we set up three different scenarios (as outlined in Sect. 4) and determine the likely increase in trading volume on lit markets due to the trading obligation for OTC trades and the BCN trading prohibition. As noted before, our data set features the intraday turnover for each single trade on a single stock basis. This special characteristic enables us to distinguish between different trade size classes. The distinction of different trade sizes within our analysis is of high importance to determine the expected turnover which will likely migrate from OTC to lit markets dependent on the probability to classify for the exemptions of the trading obligation. Due to the difficulty to predict the exact amount of volume migration as well as the actual change in fragmentation, we perform a scenario analysis.

For our subsequent analysis, we divide OTC trades and the corresponding turnover in three disjoint classes representing different probabilities to migrate to lit venues. The first class (OTC Small) represents the OTC trades below LIS[6]. The daily aggregated turnover in this category amounts to 17.70 mn € in the second quarter of 2013 for an average EURO STOXX 50 constituent. The second class (OTC Medium) includes the OTC trades between 500,000 € and 50 mn €, featuring a significantly lower probability to migrate. Trades within this category amount to the highest turnover of the three classes representing a daily average turnover of 101.15 mn € per instrument. The third class (OTC Large) incorporates the remaining OTC trades (daily average turnover of 55.25 mn € per instrument) with very high probability to remain OTC after the introduction of the trading obligation.

Overall turnover in EURO STOXX 50 constituents amounts to 1,294 billion € in the second quarter of 2013. Thereof, 48.90% were traded on lit markets, 2.64% via SIs, 0.92% in regulated dark pools (i.e. dark pools provided by RMs or MTFs) and 47.54% were conducted OTC (see Fig. 1). In our data set, BCN trades and other OTC trades are not distinguishable as they are both flagged as OTC.

---

[5] Cboe Europe operates both markets, Cboe BXE and Cboe CXE.

[6] Under MiFID I, orders that are above the LIS threshold can benefit from a waiver of pre-trade transparency. This waiver is intended to protect these orders from adverse market impact and to avoid significant price movements that can cause market distortion. For our sample of EURO STOXX 50 constituents, this LIS threshold is 500,000 €.

Since BCNs are prohibited under the new regulatory framework, it is necessary for our analysis to determine the fraction of BCN trading volume. Therefore, we rely on the market share of BCNs provided by Rosenblatt Securities, Inc. who approximate that BCN turnover accounts for 4.55% of overall trading volume in Europe.[7] Consequently, the remaining market share of OTC trading without BCN volumes amounts to 42.99%. Concerning lit venues, main markets account for 70.14% of all lit trading volume, i.e. main markets (alternative lit venues) have a market share of 34.29% (14.60%) of overall turnover in EURO STOXX 50 stocks.

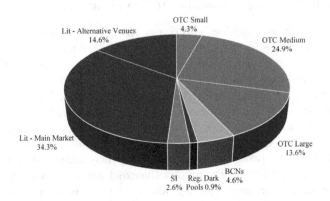

Fig. 1. Turnover distribution in EURO STOXX 50 constituents as of Q2/2013.

Due to the trading obligation introduced by MiFIR, the turnover distribution will likely change since BCNs are prohibited and since trades have to classify for certain exemptions from the trading obligation in order to still be conducted OTC. Moreover, we analyze possible changes in the European equity trading landscape based on the scenarios outlined in Sect. 4. The three trade size categories necessary to determine migrating volumes are based on the whole OTC category including BCN trades since they are not distinguishable in our sample. Therefore, we assume the trade sizes to be equally distributed in the pure OTC as well as in the BCN category. The volume migration effects for the three developed scenarios are depicted in Fig. 3 in the appendix.

In Scenario A, we assume that the OTC category as a whole including BCN volumes migrates to lit venues according to the identified trade size categories

[7] An overview of the Rosenblatt Securities, Inc. estimates for BCN market shares in Europe over time is available at
https://www.bloomberg.com/news/articles/2017-07-10/dark-pool-traders-find-mifid-workarounds-to-stay-in-the-shadows.

and their respective migration likelihood (as derived from the expert estimations). On average, this results in a volume migration of 29.16 mn € per day and instrument from OTC including BCNs to lit markets, which equals a market share increase of lit markets amounting to 7.96 % points resulting in a market share of 56.86%. The BCN volume that does not migrate to lit markets is assumed to remain OTC and the SI market share to stay constant. The remaining OTC turnover (including BCN) represents a market share of 39.58% compared to 47.54% before the trading obligation.

Scenario B differs from scenario A in the sense that SIs are able to capture the entire BCN volume so that only the OTC volume excluding the BCN volume migrates to lit venues according to the identified trade size categories and their respective migration likelihood. This results in a volume migration of 26.37 mn € per day and instrument from OTC excluding BCNs to lit markets. The market share of lit markets, therefore, increases by 7.20% points to 56.10%. Volume traded via BCNs is completely captured by SIs leading to an increase of their market share by 4.55% points to 7.19%. The remaining OTC turnover represents a market share of 35.79%.

Different to scenario B, scenario C assumes that the market share of SIs rises to a level equivalent to the US share of internalization. However, scenario C assumes the identical volume to migrate from the OTC volume excluding the BCN volumes to lit markets as scenario B. In addition, SIs (i) capture the entire BCN volumes prior to the trading obligation and (ii) are able to attract turnover from lit venues having advantages due to their possible independence from the minimum tick size regime up to the derived internalization level comparable to the US. The US volume traded in investment bank dark pools equivalent to BCNs in June 2017 accounts for 8.79% of overall turnover [35]. At the same time, the US retail internalization level is approximately 16.00% [8,35]. Due to the fact that the percentage of equity holdings relative to total financial assets for households in Europe is approximately half of the percentage of equity holdings in the US[8], we derive a predicted market share of SIs = 8.79% + 51% × 16% = 16.95% for Europe. This scenario C is assessed to be the most realistic among the participants of the survey and results in a net negative volume migration effect for lit markets of −9.38 mn € per day and instrument, leading to a (net) decrease by 2.56% points of lit markets' market share to 46.33%. SIs completely capture BCN volumes and additionally gain turnover from lit markets leading to an increase of SI's market share by 14.31% points to 16.95%. The remaining OTC turnover represents a market share of 35.79%. The resulting market shares of the different trading forms according to all three scenarios are reported in Table 1.

---

[8] Based on OECD data regarding household financial assets, equity holdings in the US represent 34.96% of a household's total financial assets while this number amounts to 17.83% in Europe, which is 51% of the US figure. The European number is a weighted mean considering only those European countries where corporations listed in the EURO STOXX 50 are headquartered.

Table 1. Market shares (Each line adds up to 100%) of the different venues and forms of trading before and after the introduction of the trading obligation.

	Lit venues	Dark pools	SI	OTC excl. BCN	BCN
Q2/2013	48.90%	0.92%	2.64%	42.99%	4.55%
Scenario A	56.86%	0.92%	2.64%	39.58%	-
Scenario B	56.10%	0.92%	7.19%	35.79%	-
Scenario C	46.33%	0.92%	16.95%	35.79%	-

Table 2 summarizes the resulting changes in trading volume and market share of lit venues according to the three scenarios. It also reports the results for the main markets and alternative venues separately. Therefore, the fraction of main market turnover of overall lit turnover is assumed to remain constant at 70.14% as in the second quarter of 2013.

Table 2. Volume migration and market share changes (average per day and instrument) of lit markets in total as well as split according to main market and alternative venues (assuming a constant fraction of main market turnover of overall lit turnover amounting to 70.14%).

	Volume Migration to Lit venues [mn €]	Change in Lit Market Shares [percentage points]	Volume Migration to Main Market [mn €]	Change in Main Market Share [percentage points]	Volume Migration to Alternative Venues [mn €]	Change in Alternative Venues' Market Share [percentage points]
Scenario A	29.16	7.96	20.45	5.58	8.71	2.38
Scenario B	26.37	7.2	18.49	5.05	7.87	2.15
Scenario C	−9.38	−2.56	−6.58	−1.8	−2.8	−0.76

5.3   Measuring the Effect of Volume Migration on Liquidity

Having determined the potential volume migration effect due to the trading obligation, this result is applied as an input to derive the impact of the new regulation on liquidity. By quantifying the actual liquidity effect due to a volume increase, we follow a similar approach as [7,27,31]. In line with [6], we follow the reasoning that additional turnover increases liquidity. According to [2], it is doubtful whether there exists a single measure that is able to capture all aspects of liquidity. In general, liquidity turns out to be a multi-dimensional variable and can therefore hardly be captured in a single one-dimensional measure. For this reason, we investigate different liquidity indicators: the relative bid-ask spread as the most common measure, Depth (10 bps) (see [9]) representing the euro volume 10 bps around the midpoint, and XLM50k as the round trip costs for trading 50,000 € (see [25]).

To determine the liquidity effect arising due to a volume increase in public limit order books, we calculate the elasticity of additional trading volume by applying an ordinary least squares regression on the above mentioned liquidity measures. For our independent variables, we rely on turnover, the inverse of the average stock price, market capitalization, the number of order book updates (tick speed), market fragmentation, and volatility. The regression equation is the following:

$$Liquidity\ measure_{i,t} = \alpha + \beta_1 \cdot Turnover_{i,t} + \beta_2 \cdot Price_{i,t}^{-1} + \beta_3 \cdot Market\ Capitalization_{i,t}$$

$$+\beta_4 \cdot Tick\ Speed_{i,t} + \beta_5 \cdot Fragmentation_{i,t} + \beta_6 \cdot Volatility_{i,t}^{MP} + \sum_{k=7}^{9} \beta_k \cdot Control_k + \epsilon_{i,t},$$

where i represents the respective constituent of the EURO STOXX 50 as of June 2013 and t is a time variable for each quarter from the first quarter of 2008 until the second quarter of 2013. $Control_k$ represents dummy variables for stocks, markets, and quarters. $\epsilon_{i,t}$ are normally distributed residuals.

To run the actual regression, we log-transform our dependent variables to make the residual a relative instead of an absolute value and to avoid unnecessary heteroscedasticity. Furthermore, this transformation results in an increase of the adjusted R^2 by more than 53% on average. To additionally avoid potential problems related to autocorrelation patterns, we aggregated the intraday observations to quarterly values. For each of the dependent variables, we run two regressions, differing in the existence of control variables. In the following, we explain why we are referring to these exogenous variables.

Since we investigate the impact of additional stock turnover on lit markets because of the trading obligation introduced by MiFIR, we include *Turnover* as the first independent variable in the regression model. Consistent with prior evidence, we include the inverse of stock prices $(Price^{-1})$ as in [27]. For low-priced stocks, the minimum tick size may constrain spreads. The inverse of the stock price captures the tendency for this effect to diminish as stock prices rise (see [20]). Moreover, we use the *Market Capitalization* of each firm to control for firm size [5]. As analysts tend to follow larger firms more closely, these stocks may feature higher trading volumes and (therefore) perhaps better liquidity indicators [30]. The variable *Tick Speed* represents the number of order book updates and can be seen as an approximation for order book activity as it measures the total number of messages submitted to the order book. As shown in [9], fragmentation has a significant impact on liquidity. To account for this effect, we rely on the inverse of the Herfindahl Hirschman Index [28], which is in line with the Fidessa

Fragmentation Index [19]. Within our setup, *Fragmentation* takes into account all lit venues and is determined on a quarterly basis for each stock separately. To account for general uncertainty, we include the order book midpoint volatility ($Volatility_{MP}$) as a measure of price deviation. In general, high volatility is associated with low liquidity and vice versa. Due to a high multicollinearity to the tick speed and turnover, we do not include the number of trades or a dummy for years in our regression setup.

5.4   Description of Regression Results

In general, the relative spread and the XLM should show consistent signs of the beta coefficients for all independent variables as visible in Table 3. In contrast to the order book depth, a high relative spread and high values for XLM indicate lower levels of liquidity, whereas higher values of order book depth indicate higher liquidity levels. The adjusted R^2 in the full models including control variables (models 1, 3 and 5) ranges from 63% to 85%. Therefore, the explanatory variables capture a high fraction of the overall variation in liquidity of European equity markets.

Under all considered models, the beta coefficients of *Turnover* are consistent. Each additional million of turnover decreases the spread by 0.038% (model 1) and 0.26% (model 2), while round trip costs measured by XLM50k are reduced by 0.14% (model 5) and 0.32% (model 6). Model 3 and 4 quantify the impact of an additional million of turnover by an increase in depth of 0.3% and 0.47%. Regarding the inverse of the stock price ($Price^{-1}$), our results almost consistently show (except model 4) that stocks with low prices are less liquid, which might (abschwächen, da nicht getestet) be explained by constraints due to minimum tick size requirements. Through higher analyst and media coverage of stocks with high market capitalization, trading interest in these stocks might be higher thus improving liquidity. However, the effect of *Market Capitalization* is only significant for the order book depth and the relative spread in model 1. The number of order book updates (*Tick Speed*) is significant for the models 2, 4, and 6. By considering the full models, the significance of this effect vanishes. Therefore, this effect seems to be captured by other variables within the full model. By investigating the effect of *Fragmentation*, we obtain consistent results (apart from model 1 and 5) and give additional empirical evidence that fragmentation improves liquidity, which is in line with the studies presented in Sect. 2. As $Volatility^{MP}$ is a measure of uncertainty, it is not surprising that high levels of stock volatility are associated with lower liquidity since liquidity providers face higher risks. Volatility as an explanatory variable is highly significant in all models. The effect of each additional basis point of volatility increases the relative spread by 5.25% (model 1) and 4.40% (model 2). XLM50k is expected to increase by 7.42% (model 5) and 7.26% (model 6). The effect on order book depth is even greater. For each additional basis point of volatility, the order book depth will decrease by 10.95% (model 3) and even up to 11.84% in model 4.

Table 3. Regression results for the EURO STOXX 50 sample based on data for the main markets. Endogenous variables are different liquidity measures. Exogenous variables are turnover, the inverse of the stock price, the market capitalization, the tick speed of the order book, fragmentation, and volatility. The full models include additional controls for stock, market, and time specific effects. We apply robust standard error estimations to correct for potential heteroscedasticity and autocorrelation biases. Please note: $*p < 0.1$, $**p < 0.05$, $***p < 0.01$.

	Dependent variable					
	log(Spread)		log(Depth10bps)		log(XLM50k)	
	(1)	(2)	(3)	(4)	(5)	(6)
Constant	−8.651	-6.884	−1.542	−1.699	1.463	2.629
	t = −36.435***	t = −46.290***	t = −7.966***	t = −12.017***	t = 7.239***	t = 17.187***
Turnover [mn]	−0.0004	−0.003	0.003	0.005	−0.001	−0.003
	t = −1.239	t = −10.525***	t = 11.144***	t = 15.788***	t = −4.532***	t = −12.737***
$Price^{-1}$	1.633	1.4	−0.97	0.287	2.002	1.183
	t = 4.791***	t = 7.857***	t = −3.180***	t = 2.272**	t = 5.312***	t = 6.048***
MarketCap [bn]	−0.003	−0.0003	0.011	0.006	−0.002	0.0002
	t = −1.760*	t = −0.269	t = 5.330***	t = 6.658***	t = −1.160	t = 0.178
Tick Speed [k]	0.0003	0.002	0.0001	−0.001	0.0001	0.002
	t = 1.109	t = 9.889***	t = 0.499	t = −3.257***	t = 0.377	t = 8.763***
Fragmentation	0.321	−0.187	0.533	0.705	0.149	−0.157
	t = 3.638***	t = −2.431**	t = 8.339***	t = 12.237***	t = 2.281**	t = −2.086**
$Volatility^{MP}$	524.91	440.217	−1,095.38	−1,183.61	742.487	726.289
	t = 7.264***	t = 6.965***	t = −19.707***	t = −18.092***	t = 11.660***	t = 12.839***
Controls:						
Stocks	Yes		Yes		Yes	
Marketplace	Yes		Yes		Yes	
Quarters	Yes		Yes		Yes	
Observations			1,034			
R^2	0.654	0.348	0.858	0.725	0.713	0.417
Adjusted R^2	0.634	0.345	0.851	0.724	0.697	0.414
Residual Std. Error	0.417 (df = 978)	0.559 (df = 1027)	0.331 (df = 978)	0.45 (df = 1027)	0.392 (df = 978)	0.546 (df = 1027)
Max VIF	4.88	2.14	4.88	2.14	4.88	2.14
Mean VIF	2.37	1.65	2.37	1.65	2.37	1.65

5.5   Quantifying the Effect on Liquidity

In a final step, we combine the preceding analyses and estimate the liquidity effects due to changes in trading volume. Since the regression analysis of turnover and liquidity is based on data from the main markets, we calculate the liquidity effects for the main lit markets. However, the liquidity effect for the alternative lit markets should be of similar magnitude. For the analysis, we use the determined volume migrations to main markets of 20.45 mn € (scenario A), 18.49 mn € (scenario B) as well as −6.58 mn € (scenario C) as shown in Table 2 and combine them with the beta coefficient of the full models (1, 3 and 5) representing the liquidity sensitivity. Because the beta coefficients are point estimates and we do not neglect their variability, we also compute the 95% confidence interval represented by error bars in Fig. 2, which summarizes our results regarding the effect of the trading obligation on main market liquidity.

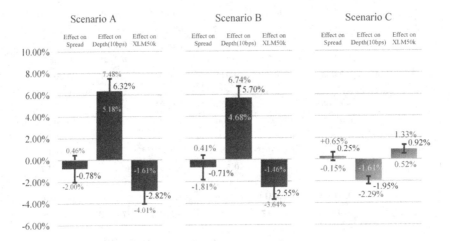

Fig. 2. Effect of turnover variations on liquidity on the main market measured by spread, Depth(10 bps), and XLM50k.

Assuming that the trading obligation changes the European market land-scape according to scenario A and taking the full regression models as the model of choice, the effect of additional turnover amounting to 20.45 mn € per stock and day (see Table 2) will lower the relative spread on the main market by 0.78%. The effect on XLM50k is slightly higher. Due to the additional turnover on the main market, XLM50k will decrease by 2.82%. In terms of order book depth, liquidity will increase by 6.32%.

The impact on liquidity measures in scenario B is slightly lower than the effect within scenario A due to the assumption that BCNs will completely migrate to SIs. The additional volume of 18.49 mn € per stock and day on the main market results in a decrease of the relative spread of 0.71%. XLM50k reduces by 2.55% and order book depth increases by 5.70%. Consequently, the trading obligation introduced by MiFIR increases liquidity on the main market in scenarios A and B.

In scenario C, OTC volumes still migrate to the main market but as SIs gain competitive advantages and, together with the BCN volumes, more than double their volumes relative to pre-MiFID II levels, the net effect of volume migration on the main market is negative (-6.58 mn € per stock and day). Therefore, the relative spread on the main market increases by 0.25%. While the XLM50k increases by 0.92%, the effect on order book depth is even higher as it decreases by 1.95%. Therefore, scenario C reveals that the trading obligation might also lead to decreasing liquidity on main markets due to increased trading via platforms of SIs.

In order to equip market participants with a toolbox to assess changes in market shares of venues and liquidity based on their own expectations, we conduct a sensitivity analysis in Sect. A.2 of the appendix with respect to different thresholds regarding the migration of OTC volumes to lit markets for each of the three scenarios.

6   Discussion and Limitations

In contrast with many retrospective studies on the impact of a new regulatory regime on liquidity (e.g., [9, 26]), this study aims at quantifying the impact of new regulation on market liquidity prospectively. Specifically, we investigate the potential impact of the trading obligation for shares introduced by MiFID II/MiFIR, which is applicable since January 2018. The trading obligation represents a cornerstone of the revised European financial market regulation to achieve the regulatory goal of increased transparency in financial markets by bringing OTC trades that do not classify for certain exemptions to organized trading venues and SIs. Due to the trading obligation for shares, OTC trading is restricted by MiFID II/MiFIR leading to a likely volume migration to organized venues. Based on a scenario analysis, we determine possible outcomes of this volume migration from the OTC market to organized venues and analyze the effect of additional turnover on main market liquidity.

Our scenarios aim to capture possible long-term outcomes of MiFID II/ MiFIR for future European equity trading. On the one hand, BCNs are prohibited under the new regulatory regime so BCN turnover has to move to other forms of trading. On the other hand, small and medium sized OTC trades will - at least to a certain percentage - move to organized venues due to the trading obligation introduced by MiFIR. Scenarios A and B reflect this volume migration differing only in the assumption that SIs are able to capture the entire BCN volume in scenario B. In both scenarios, more volume is executed on lit venues thereby increasing liquidity. Scenario C, which the survey participants assessed to be the most realistic one, describes the possibility that SIs might gain competitive advantages through the trading obligation in combination with possible minimum tick size exemptions and best execution requirements. In this scenario, we assume that SI market shares rise towards the level of internalization in the US (adjusted to the European context) leading to decreasing trading volumes and liquidity on lit venues. Scenario C therefore reveals that the trading obligation in combination with other MiFID II/MiFIR rules might also lead to negative effects for lit markets and public price discovery.

Besides an effect on investors' transaction costs, the trading obligation for shares also has an impact on the European economy as a whole. As liquidity is a priced risk factor [1, 33], increased liquidity on lit venues due to higher trading volumes lowers the cost of capital for issuers thereby having a positive effect on the real economy in Europe.

We are aware that our study has some limitations. Since this study aims to predict consequences of MiFID II/MiFIR, it should be validated once the effects of the new regime have fully unfolded and sufficient data after the introduction is available, e.g. in 2019. Furthermore, this study can be seen as a benchmark for studies analyzing the actual impact of the trading obligation. From a methodological point of view, our predictive analysis is based on different scenarios that only represent possible future outcomes although they build on reasonable assumptions. Moreover, we set up different regression models with and without control variables to assess the impact of additional turnover on market liquidity leading to almost entirely consistent results. Although our model already explains a great fraction of the overall variation (up to 85%), there is still the chance that we missed essential variables explaining further variation in the endogenous liquidity measures. The data set used to derive our results offers the advantage to analyze possible volume shifts between different trading categories in great detail since it covers all forms of trading for the most liquid European stocks (main market, other lit markets, SIs, dark pools, OTC). The data set already ends in June 2013. However, market share distribution and overall trading volume in June 2013 are highly comparable to June 2017. Additionally, a more recent data set would suffer from possible announcements effects of MiFID II/MiFIR and the effects of the new trading mechanisms that were already implemented in light of the new regulation.

Furthermore, our estimates regarding trading volumes and liquidity after the introduction of the trading obligation for shares do not consider relevant but uncertain future changes in European markets such as the Brexit and the resulting potential classification of the United Kingdom as a third country from the perspective of the European Union. While this event undoubtedly will have a significant impact on European financial markets, its effect on trading volumes, liquidity, and market quality cannot be reliably reflected yet since it is also uncertain how market participants and trading venues will react before and after the Brexit.

7 Conclusion

The trading obligation imposed by MiFID II/MiFIR, which has to be applied since January 2018, requires all trades in shares to be executed on Regulated Markets, Multilateral Trading Facilities, or Systematic Internalisers unless predefined criteria apply. According to our predictive scenario analysis, the trading obligation for shares might increase transparency, trading volume, and liquidity on lit markets. In this study, we investigate the potential impact of the trading obligation imposed by MiFIR on liquidity of EURO STOXX 50 constituents. Drawing on a comprehensive data set and three scenarios, we are able to show that the trading obligation might increase turnover and liquidity on lit venues. Specifically, relative bid-ask spreads might decrease by 0.71–0.78%, round trip transaction costs for 50,000 € by 2.55–2.82%, and order book depth 10 bps around the midpoint might increase by 5.70–6.32% depending on the respective scenario.

However, there is also the possibility that Systematic Internalisers gain competitive advantages through the trading obligation in combination with their possible independence from the minimum tick size regime, which might even lead to decreasing turnover and liquidity on lit venues. In our online survey among industry experts, which we conducted in order to prospectively capture the market's belief about the most probable outcome, the respondents indicated this scenario is the most realistic one. According to this scenario, spreads might increase by 0.25%, round trip transaction costs of 50,000 € by 0.92% and market depth 10 bps around the midpoint might decrease by 1.95% on lit venues. This effect on liquidity triggered by the MiFIR trading obligation not only increases trading costs for investors in European equities trading, but also has a negative impact on issuers due to higher cost of capital and thereby on the real economy in Europe.

As this study is forward-looking, future research should evaluate the actual effect of the trading obligation for shares on trading volume and liquidity once sufficient observation time after the application of the regulation is available. Our results may serve as a benchmark for the evaluation of the trading obligation for shares in future studies. Moreover, the effect of new transparency requirements in MiFID II/MiFIR on market quality, the market share distribution in non-equities after the introduction of OTFs and the trading obligation for derivatives are areas worth to be studied in order to evaluate the consequences of the new regulatory regime MiFID II/MiFIR.

A   Appendix

A.1   Volume Migration Scenarios

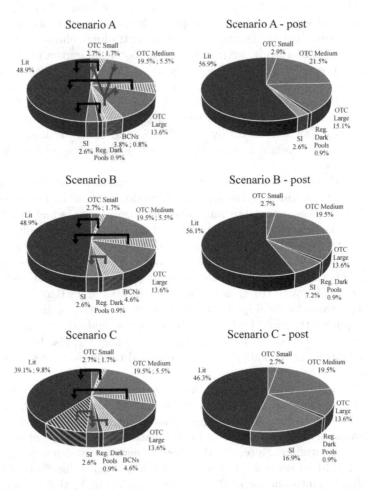

Fig. 3. Effect of three different scenarios on the European equity trading landscape. Striped areas represent market shares that migrate to other trading categories. Numbers on the left indicate the percentage points that remain in the respective trading category, while numbers on the right represent the percentage points that move to other categories. If there is only one number, then the whole category either remains or migrates completely. The changes in market share and resulting trading volumes depicted in this figure are based on the assumption that 39% of the OTC trades smaller than 500,000 €, 22% of the OTC trades between 500,000 € and 50 mn €, and none of the OTC trades larger than 50 mn € will migrate to lit markets (according to the results of an online survey among 111 industry experts).

A.2   Sensitivity Analysis

Having identified the effects of the trading obligation on market share distribution in European equity trading and on main market liquidity according to the three scenarios, this section provides the results of a sensitivity analysis with respect to the chosen volume migration thresholds for the identified OTC trade size categories. Since large trades are highly likely to classify for the exemptions of the trading obligation, we keep the threshold of the "OTC Large" category constant at 0% for the sensitivity analysis since this assumption is supported by 98% of the survey respondents. The migration threshold of the "OTC Small" ("OTC Medium") category is varied between 15% and 60% (5% and 45%) for sensitivity purposes. Table 4 shows the results of the sensitivity analysis for the effect of the trading obligation on changes in market shares of lit markets. The shaded area ("39/22/0") of Tables 4 and 5 shows the estimated migration of OTC volumes in the different size categories according to our survey among market experts. In all cases, scenarios A and B lead to positive effects of the trading obligation for lit markets increasing turnover in EURO STOXX 50 constituents relative to other trading categories. Scenario C, however, only leads to a rising market share of lit markets if the fraction of migrating volume from the small and medium OTC trade size category is large enough, i.e. more than 45% of the "OTC Small" and more than 30% of the "OTC Medium" category, to compensate the volume migration from lit markets to SIs. In all other cases of scenario C, the trading obligation leads to declining market shares of lit markets.

Table 4. Net effect of the trading obligation on lit markets regarding changes in market share. The results are based on varying migration thresholds of the three identified OTC trade size categories (OTC Small/OTC Medium/OTC Large). The shaded area highlights the results of the chosen values for the scenario analysis conducted in the previous section.

				Changes in Market Share				
60/45/0	55/40/0	50/35/0	45/30/0	39/22/0	35/20/0	30/15/0	25/10/0	15/5/0
Scenario A 15.30%	13.70%	12.10%	10.50%	8.00%	7.20%	5.60%	4.00%	2.10%
Scenario B 13.90%	12.40%	10.90%	9.50%	7.20%	6.50%	5.10%	3.60%	1.90%
Scenario C 4.10%	2.60%	1.20%	-0.30%	-2.60%	-3.20%	-4.70%	-6.20%	-7.90%

The varying amount of migrating volume to lit markets also leads to different effects on main market liquidity. Table 5 reports the results of the sensitivity analysis with varying migration thresholds of the OTC trade size categories for the three different liquidity measures spread, Depth(10 bps) and XLM50k on the main market. Similar to the effect on market shares of lit markets, main market liquidity increases in both scenarios A and B indicated by decreasing spreads and round trip transaction costs measured by XLM50k. Also, order book depth increases. In scenario C, however, liquidity on the main market decreases for most depicted OTC migration thresholds and only increases if the fraction of migrating small and medium sized OTC trades to lit markets is large enough.

Table 5. Net effect of the trading obligation on main market liquidity. The results are based on varying migration thresholds of the three identified OTC trade size categories (OTC Small/OTC Medium/OTC Large). The shaded area highlights the results of the chosen values for the scenario analysis conducted in the previous section.

Spread Change

	60/45/0	55/40/0	50/35/0	45/30/0	39/22/0	35/20/0	30/15/0	25/10/0	15/5/0
Scenario A	-1.50%	-1.30%	-1.20%	-1.00%	-0.80%	-0.70%	-0.50%	-0.40%	-0.20%
Scenario B	-1.40%	-1.20%	-1.10%	-0.90%	-0.70%	-0.60%	-0.50%	-0.40%	-0.20%
Scenario C	-0.40%	-0.30%	-0.10%	0.00%	0.30%	0.30%	0.50%	0.60%	0.80%

Effect on Depth(10bps)

	60/45/0	55/40/0	50/35/0	45/30/0	39/22/0	35/20/0	30/15/0	25/10/0	15/5/0
Scenario A	12.50%	11.10%	9.80%	8.40%	6.30%	5.70%	4.40%	3.10%	1.60%
Scenario B	11.30%	10.00%	8.80%	7.60%	5.70%	5.20%	4.00%	2.80%	1.50%
Scenario C	3.20%	2.00%	0.90%	-0.20%	-2.00%	-2.50%	-3.60%	-4.60%	-5.90%

XLM50k Change

	60/45/0	55/40/0	50/35/0	45/30/0	39/22/0	35/20/0	30/15/0	25/10/0	15/5/0
Scenario A	-5.40%	-4.80%	-4.20%	-3.70%	-2.80%	-2.60%	-2.00%	-1.40%	-0.80%
Scenario B	-4.90%	-4.40%	-3.90%	-3.30%	-2.60%	-2.30%	-1.80%	-1.30%	-0.70%
Scenario C	-1.50%	-0.90%	-0.40%	0.10%	0.90%	1.20%	1.70%	2.20%	2.90%

References

1. Acharya, V., Pedersen, L.: Asset pricing with liquidity risk. J. Financ. Econ. **77**(2), 375–410 (2005)
2. Amihud, Y.: Illiquidity and stock returns: cross-section and time-series effects. J. Financ. Markets **5**(1), 31–56 (2002)
3. Brauers, J., Weber, M.: A new method of scenario analysis for strategic planning. J. Forecast. **7**(1), 31–47 (1988)
4. CESR: CESR technical advice to the European Commission in the context of the MiFID Review - equity markets. CESR/10-394 (2010). https://www.esma.europa.eu/sites/default/files/library/2015/11/10_394.pdf
5. Chordia, T., Roll, R., Subrahmanyam, A.: Co-movements in bid-ask spreads and market depth. Financ. Anal. J. **56**(5), 23–27 (2000)
6. Chordia, T., Roll, R., Subrahmanyam, A.: Recent trends in trading activity and market quality. J. Financ. Econ. **101**(2), 243–263 (2011)

7. Christensen, H.B., Hail, L., Leuz, C.: Capital-market effects of securities regulation: prior conditions, implementation, and enforcement. Rev. Financ. Stud. **29**(11), 2885–2924 (2016)
8. Comerton-Forde, C., Malinova, K., Park, A.: Regulating dark trading: Order flow segmentation and market quality. Working Paper (2016)
9. Degryse, H., de Jong, F., van Kervel, V.: The impact of dark trading and visible fragmentation on market quality. Rev. Financ. **19**(4), 1587–1622 (2015)
10. European Commission: IP/02/1706 press release: Investment services: Proposed new Directive would protect investors and help investment firms operate EU-wide (2002). http://europa.eu/rapid/press-release_IP-02-1706_en.htm
11. European Commission: Proposal for a Directive of the European Parliament and of the Council on markets in financial instruments repealing Directive 2004/39/EC of the European Parliament and of the Council (2011). http://ec.europa.eu/internal_market/securities/docs/isd/mifid/COM_2011_656_en.pdf
12. European Commission: Commission Delegated Regulation (EU) .../... of 28.8.2017 amending Delegated Regulation (EU) 2017/565 as regards the specification of the definition of systematic internalisers for the purposes of Directive 2014/65/EU (2017). http://ec.europa.eu/info/law/better-regulation/initiative/109985/attachment/090166e5b4ac6f24_en
13. European Commission: Commission Delegated Regulation (EU) 2017/588 supplementing Directive 2014/65/EU of the European Parliament and of the Council with regard to the regulatory technical standards on the tick size regime for shares, depositary receipts and exchange-traded funds (2017). http://eur-lex.europa.eu/legal-content/EN/TXT/PDF/?uri=CELEX:32017R0588&from=en
14. European Parliament and Council: Directive 2004/39/EC of 21 April 2004 on markets in financial instruments (MiFID I) (2004). http://eur-lex.europa.eu/legal-content/EN/TXT/PDF/?uri=CELEX:32004L0039&qid=1494934175734&from=EN
15. European Parliament and Council: Directive 2014/65/EU of 15 May 2014 on markets in financial instruments (recast) (MiFID II) (2014). http://eur-lex.europa.eu/legal-content/EN/TXT/PDF/?uri=CELEX:32014L0065&from=EN
16. European Parliament and Council: Regulation (EU) No 600/2014 of 15 May 2014 on markets in financial instruments (MiFIR) (2014). http://eur-lex.europa.eu/legal-content/EN/TXT/PDF/?uri=CELEX:32014R0600&from=EN
17. Federation of European Securities Exchanges: European equity market report (2017). http://www.fese.eu/statistics-market-research/european-equity-market-report
18. Federation of European Securities Exchanges: Tick size regimes (2017). http://www.fese.eu/statistics-market-research/tick-size-regimes
19. Fidessa: Fidessa fragmentation index: Making sense of global fragmentation (2017). http://fragmentation.fidessa.com/
20. Flannery, M.J., Kwan, S.H., Nimalendran, M.: Market evidence on the opaqueness of banking firms' assets. J. Financ. Econ. **71**(3), 419–460 (2004)
21. Gomber, P., Clapham, B., Lausen, J., Panz, S.: The impact of MiFID II/MiFIR on european market structure: a survey among market experts. J. Trading **13**(2), 35–46 (2018)
22. Gomber, P., Jäger, B.: MiFID: Eine systematische Analyse der Zielerreichung. Zeitschrift für Bankrecht und Bankwirtschaft **16**(1), 40–53 (2014)
23. Gomber, P., Sagade, S., Theissen, E., Weber, M.C., Westheide, C.: The state of play in European over-the-counter equities. J. Trading **10**(2), 23–32 (2015)

24. Gomber, P., Sagade, S., Theissen, E., Weber, M.C., Westheide, C.: Spoilt for choice: Order routing decisions in fragmented equity markets. Working Paper (2016)
25. Gomber, P., Schweickert, U., Theissen, E.: Liquidity dynamics in an electronic open limit order book: an event study approach. Eur. Financ. Manage. **21**(1), 52–78 (2015)
26. Gresse, C.: Effects of lit and dark market fragmentation on liquidity. J. Financ. Markets **35**, 1–20 (2017)
27. Hendershott, T., Charles, J.M., Menkveld, A.J.: Does algorithmic trading improve liquidity? J. Finance **66**(11), 1–33 (2011)
28. Hirschman, A.O.: The paternity of an index. Am. Econ. Rev. **54**(5), 761–762 (1964)
29. Huss, W.R.: A move toward scenario analysis. Int. J. Forecast. **4**(3), 377–388 (1988)
30. Lambert, R.A., Leuz, C., Verrecchia, R.E.: Accounting information, disclosure, and the cost of capital. J. Account. Res. **45**(2), 385–420 (2007)
31. O'Hara, M., Ye, M.: Is market fragmentation harming market quality? J. Financ. Econ. **100**(3), 459–474 (2011)
32. Oxera Consutling Ltd.: Monitoring prices, costs and volumes of trading and post-trading services: Report prepared for European Commission DG Internal Market and Services: (MARKT/2007/02/G)
33. Pástor, Ľ., Stambaugh, R.F.: Liquidity risk and expected stock returns. J. Polit. Econ. **111**(3), 642–685 (2003)
34. Postma, T.J., Liebl, F.: How to improve scenario analysis as a strategic management tool? Technol. Forecast. Soc. Change **72**(2), 161–173 (2005)
35. Rosenblatt: Let there be light: Rosenblatt's monthly dark liquidity tracker (2017). http://rblt.com/lettherebelight_details.aspx?id=715
36. Schoemaker, P.J.H.: Scenario planning: a tool for strategic thinking. Sloan Manage. Rev. **36**(2), 25–40 (1995). https://search.proquest.com/docview/1302991850?accountid=10957

Open Innovation Effectiveness in the Financial Services Sector

Francesco Piobbici[(⊠)], Federico Rajola, and Chiara Frigerio

Università Cattolica del Sacro Cuore, Milan, Italy
Francesco.piobbici@gmail.com

1 Introduction

Many factors are changing the financial services industry. At first, we have to consider the macroeconomic scenario characterized by low interest rates, leading to a reduction of the institutions' profitability and promoting investments aimed to increase the organizations' efficiency. The regulatory framework is another element to consider, since, after the financial crisis, the authorities introduced new norms, some of them are currently under implementation, like the Market in Financial Instruments Directive II, the Insurance Distribution Directive or the General Data Protection Regulation. The last element we have to consider is the technological one: during the years, many innovations changed the customers' behaviors, introducing new needs that institutions have to take into account in their strategic plans. Google, Apple, Facebook and Amazon, also called GAFA, have introduced new ways to interact with consumers that are based on instant and easy access to different functionalities. The success of those new models lead the so-called GAFA to became an attractive alternative to the traditional financial institutions, considering also their propensity to enter in this market (Accenture 2017). The implementation of new technologies helps institutions to face this issue; latest application of artificial intelligence to the business processes of the financial sector are gaining much attention, since they enable institutions to be more effective in meeting the customers' needs (Raconteur 2018). As an example, we could consider the use of machine learning algorithms in order to choose the right portfolio allocation for customers. This technology has been implemented in some mobile applications[1], creating a digital financial advisor that can manage low amount of savings. Those services has been quite successful since people that would like to save money use them, but they are not able to pay the fees of private bankers. Moreover, there are new companies called fintech that are getting into this sector with new services based on innovative technologies. The growing investments on the new entrants demonstrate their relevance for the sector: at the end of the 2016, they reached the amount of 13.6 billion of US Dollars (KPMG 2017). Given the broad diffusion of successful innovations in the market, institutions embraced open innovation strategies

---

[1] Moneyfarm is an example of this business model's type. It is an Italian start-up founded in the 2011, that is offering robo-advisory services both in Italy and in UK. For more information read Financial Times (2018).

© Springer Nature Switzerland AG 2019
N. Mehandjiev and B. Saadouni (Eds.): FinanceCom 2018, LNBIP 345, pp. 27–38, 2019.
https://doi.org/10.1007/978-3-030-19037-8_2

in order to capture ideas developed outside the company and implementing them in their business models. The aim of this study is to analyze some of the choices that an institution could take in order to open its boundaries, understanding which is the more effective through a consistent data set, built on a panel of Italian financial institutions.

2 Literature Review

2.1 Innovation in the Financial Services Sector

Innovation in the financial services sector has been widely studied in the past, considering both services evolution issues (Barras 1990) and product innovation ones (Boot and Thakor 1997). This topic became more relevant after the 2007 financial crisis, since policies aimed to accelerate innovation were implemented without a complete evaluation of their effects, leading to negative consequences for the whole system: product innovation related to the creation of collateralized debt obligations leads to a broad distribution of bad quality products to investors (Sveiby 2012). The financial crisis showed that institutions' innovation processes could have dangerous effects on the society, promoting more research on how to manage those processes and taking into account all the elements that characterize the complexity of the financial sector (Lerner and Tufano 2011). This issue is still relevant, since innovation in the financial services could allow institutions to face the market dynamism, characterized by the increasing competition of new players and increasing complexity in customer needs (Khraisha and Arthur 2018). Moreover, innovation in the financial services sector needs further studies since there are many issues related to barriers to innovation that needs to be overcome (Das et al. 2018). Financial innovation has been widely studied and theorized during the years (Mention and Torkkeli 2012); in this paper we are going to consider the definition by Frame and White (2004), that groups financial innovation as new products, new services, new production processes and new organizational forms.

2.2 Open Innovation and Its Relevance for the Financial Sector

Open innovation is defined as "*a paradigm that assumes that a firm can and should use external ideas as well as internal ideas, and internal and external paths to market, as they look to advantage their technology*" (Chesbrough et al. 2006). Hence, according to this theory firms could externalize a part of their R&D function to third parties, in order to get more innovative ideas. This paradigm has been widely studied by literature, focusing on different elements: Sieg et al. (2010) studied the processes of problem foundation and problem solving of the R&D unit when it is integrated with external experts; Chatenier et al. (2010) developed a competence profile that the members of the R&D team should have in order to be able to integrate external innovative ideas; Almirall and Masanell (2010) developed a model about how to balance the benefits from discovery new ideas and the cost of the divergence with the external resources. This theory is also been widely analyzed through the lens of absorptive capacity, that is considered a key element for a successful implementation of OI strategies (West and Bogers 2014).

However little attention is given to the financial services industry, even if OI is becoming a relevant issue for this sector, since institutions are cooperating with other firms and with users along the innovation processes (Mention and Torkkeli 2012). Hence, the impact of the OI framework in the financial services industry has to be analyzed more in depth and new studies should be conducted using large samples, in order to get a better explicative power (Salampasis et al. 2014). Then recent studies on financial services innovation show that the main players that bring innovations to the financial services are not institutions (Gomber et al. 2018). Moreover, the OI framework could help understanding how to encompass effectively the external sources of innovation.

3 Research Question and Hypotheses

Considering the current attention on the topics that concern innovation for the financial services industries, and the need of further studies that investigate on open innovation strategies of financial institutions, the research question that we would like to answer in this paper is: which is the best type of collaboration in order to make the open innovation strategies of financial institutions be effective?

Since OI is a wide research domain and literature has identified numerous mechanisms that a firm could use in order to implement OI strategies (West and Bogers 2014; Stanko et al. 2017), we have chosen the one referred to co-operations aimed to develop innovation related activities, given the grooving interest in this field (Belderbos et al. 2017; Kaupilla 2015). This approach will allow us to capture the heterogeneity between different partnership typologies, identifying which one is more productive in terms of innovation outcomes. Indeed the aim of this analysis is not to find if there is a positive correlation between collaborations and innovative performance, but it is to understand which is the most effective type of collaboration. Hence, we included in the analysis the four types of co-operation practices identified by Mention (2011); they are:

- inter-firm collaborations with other firms of the group in order to spread ideas generated in one part of the institution to all the organization;
- market-based collaborations, based on the involvement of suppliers and users in the development process of innovation. In this way the firm could exploit advanced instrument supplied by other firms, in order to create product and services based on the costumers' requirements;
- collaborations with competitors;
- science-based collaborations, based on partnerships with universities or higher education institutions, Government and public or private research institutes.

Moreover, we are going to answer to the previous research question, testing the following hypotheses.

H1. Financial institutions involved in inter-firm collaborations are more likely to have introduced product or service (both radical and incremental), process or organizational innovations, compared to those not involved in inter-firm collaborations.

H2. Financial institutions involved in market-based collaborations are more likely to to have introduced product or service (both radical and incremental), process or organizational innovations, compared to those not involved in market-based collaborations.

H3. Financial institutions involved in collaborations with competitor firms are more likely to have introduced product or service (both radical and incremental), process or organizational innovations, compared to those not involved in collaborations with competitors.

H4. Financial institutions involved in science-based collaborations are more likely to have introduced product or service (both radical and incremental), process or organizational innovations, compared to those not involved in science-based collaborations.

4   Methodology

4.1   Dataset: Community Innovation Survey

The Community Innovation Survey is conducted by Eurostat and it aims to map the innovative activities of European enterprises. The CIS has been used in order to empirically test relations that concern the OI innovation framework, since it offers a wide set of items that regard co-operation practices and knowledge sources used by firms in their innovative activities. Laursen and Salter (2006) was one of the first main contributions in OI literature that used the CIS in order to the test the relation between firm's knowledge sources and its innovative performance. Furthermore, many authors uses the CIS data set for their analysis, like Garriga et al. (2013) and Mention (2011). This data set could be very useful in the development of the analysis, since it covers a wide range of the Italian financial institutions; hence, it allows testing the hypotheses using a large number of observations that are difficult to collect using other methodologies. Moreover, the data set includes observations related to other sectors and further studies could be conducted, observing and comparing the differences between sectors.

The data set used in the development of the paper derives from the CIS conducted in an Italian panel of both manufacturing and service firms, during the two-year period 2012–2014. In this case, we have extrapolated the sub-sample of firms involved in the financial services sector. The panel is composed by 845 observations, 67% of them are financial institutions like banks, 22% are insurance, re-insurance companies and health funds, while 11% are auxiliary activities.[2]

---

[2] More details of the sample are reported in Fig. 1.

4.2 Measures

In order to test the hypothesis, we run a similar model of the one presented by Mention (2011), where it was analyzed if co-operation activities and information sources effect innovation performances of services firms. The following model differs from the Mention's one because the variables that measure the knowledge sources are dropped out, since the CIS2014 does not present questions on this argument. In addition the extended model includes variables that increase its predictive power, that are described later on, since they encompass other elements that helps to explain firms' innovation performances.

The model is applied to the dependent variables referred to the classification that describes the types four of innovations for the financial sector, presented by Frame and White (2004). However, product and service innovations are grouped together, in order to allow us to focus our analysis also in the different effect of OI strategies for what concerns radical and incremental innovation. Hence, the effectiveness of OI strategies is measured considering if the institution has introduced one or more of the innovation types considered. Moreover, we are going to run 5 models, one for each type of innovation highlited.

Here all the variables included in the regressions are listed.

Dependent Variables

- PSINN: 1 if the firm has introduced a product or service innovation during the two year period covered by the survey.
- NEWMKT: 1 if the firm has introduced a product or service innovation that is new or significantly improved for the market. It measures radical innovation performance.
- NEWFIRM: 1 if the firm has introduced a product or service innovation that is new or significantly improved for the firm, but it is already available in the market. It measures incremental innovation performance.
- PRCINN: 1 if the firm has introduced a process innovation during the two year period covered by the survey. The activities included in this variable are: new or significantly improved methods of producing goods or services, deliver or distribution methods, supporting activities for the processes.
- ORGINN: if the firm has introduced a process innovation during the two year period covered by the survey. The activities included in this variable are: new business practices for organizing procedures, new methods of organizing work responsibilities and decision making and new methods of organizing external relations with other firms and public institutions.

Predictors

- COIF: 1 if the firm is involved in collaborations with other firms of the group.
- COMKT: 1 if the firm is involved in market-based collaborations.
- COCPT: 1 if the firm is involved in co-operation activities with competitors.
- COSCN: 1 if the firm is involved in science-based collaborations.

Control Variables

Considering the empirical analysis of Laursen and Salter (2006) and Cheng and Shiu (2014), we are going to control the effect of firms' size, measured by the number of employees (emp14), since larger firms own more resources and they are able to develop more projects. Moreover, we are going to consider also the economic performance of the firms, measured by the sales' income (turn14). The model considers also the effect of firm's R&D policies, controlling if the firm uses an internal unit for this purpose (RRDIN) or if it externalizes this function to other partners (RRDEX).

5 Main Results

Before proceeding with the description of the regressions' results, we are going to describe briefly some evidence that arise from the descriptive statistics that are reported on Fig. 1. During the period covered by the survey, Italian financial institutions were involved in different innovative activities: the 40% of them introduced product or service innovation, the 19% of them were new to the market while the 31% new to the firm. This means that institutions were not only involved in incremental innovation process, developing ideas already introduced in the market, but also they were also involved in radical innovation initiatives, even if it is not usual for this sector (Das et al. 2018). Moreover, the 37% of the sample developed process innovations, while the 52% introduced organizational innovation. Hence, Italian financial institutions are introducing innovations that could affect different parts of their value chain, from the product design and development to the services' distribution.

For what concerns collaborations, we can notice that a part of the sample is involved in those type of activities aimed to develop innovations. Collaborations with suppliers and consumers are the most used, since the 10% of the respondents established them. Science-based collaboration are the second most used (8%), followed by intra-firms collaborations (6%) and collaborations with competitors (4%).

Looking at the outputs of the regressions, reported in Table 1, we can notice that both the Cox & Snell R-square and the Negerlkerke R-square[3] have a positive value, hence the model could describe part of the variation in the innovative performances of the institutions that belong to the sample.

Considering the model classification[4], we can state that the models have a good predictive power: in most of the innovation types the models could predict more than 70% of the correct outcome, except for the organizational innovation case, where the

---

[3] They are two pseudo R-square values and they indicate approximately how much the variation in the outcome is explained by the model, since they are built differently with respect of R-square measures of OLS regressions.

[4] Model classification is a measure that allows to test the prediction generated by the model, classifying the probability that the model could generate correct predicted categories based on the values of the predictor variables. The cut value is set to 0.5, hence we plotted in Fig. 3 the ROC curves of the regressions. We can notice that areas under the ROC curves are higher than 0.5 in all the regressions, proving the good predictive power of the model.

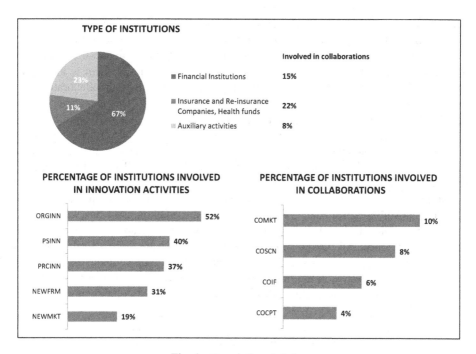

Fig. 1. Descriptive statistics

Table 1. Regressions' outputs

	PSINN		PRCINN		ORGINN		NEWMKT		NEWFRM	
	Beta	Odds-ratio	Beta	Odds-ratio	Beta	Odds-ratio	Beta	Odds-ratio	Beta	Odds-ratio
COIF	−0,13	0,87	−0,27	0,76	−0,25	0,8	0,28	1,32	0,29	1,34
COMKT	1,26**	3,54	1,54**	4,67	0,93**	2,5	0,63	1,87	0,91**	2,47
COCPT	0,74	2,10	0,44	1,55	0,85	2,4	0,35	1,41	0,64	1,9
COSCN	0,02	1,02	0,40	1,5	0,47	1,6	0,39	1,48	−0,3	0,74
RRDIN	2,41**	11,14	1,53**	4,6	0,81**	2,25	1,36**	3,9	1,64**	5,17
RRDEX	0,97	2,63	1,34*	3,81	−0,12	0,89	0,8	2,22	0,76	2,13
turn14	0	1	0	1	0	1	0	1	0	1
emp14	0	1	0	1	0	1	0	1	0	1
Constant	−0,99**	0,37	−0,93**	0,4	−0,37**	0,69	−1,93**	0,15	−1,25**	0,29
Cox & Snell R Sq.	0,19		0,14		0,12		0,11		0,13	
Nagelkerke R Sq.	0,26		0,19		0,15		0,18		0,19	
Model classification	73°%		73%		65%		82%		75%	

** Sig. at 1% level
* Sig. at 5% level

percentage is equal to 65%. Moreover, establishing collaborations with other players could help significantly financial institutions to be effective in their innovative activities.

The outputs of the regressions show us that, even if Italian financial institutions were involved in different typology of collaborations, only one of them is the more effective. In all the regressions, the variable referred to market-based collaborations is the only one to be highly significant with respect of the others. This is not true in the case product or service innovation that are new to the market, this may be due to the fact that this type of innovation is the less pursued in the financial services industries with respect to the incremental one (Das et al. 2018). The descriptive statistics reported in Fig. 1 give us evidence of this statement, since only a small part of the sample introduced products or services that are new to the market.

Moreover, we can notice that this variable has a different effect on the different types of innovation performance. We have plotted the odds-ratios of this variable for each kind of innovation in Fig. 2, except for the radical innovation case, in order to show this difference. All the odds-ratios reported in Fig. 2 are higher than 2 and this means that financial institutions involved in collaborations with suppliers or users are more likely to be effective in the introduction of an innovation, whichever is the type, with respect of institutions that are not involved in this type of collaborations.

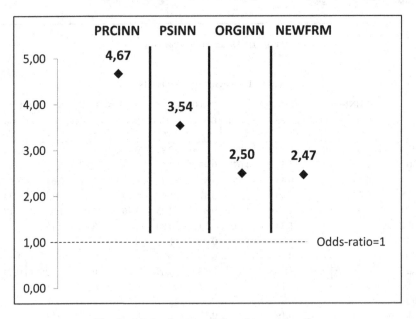

Fig. 2. Market-based collaborations odds-ratios

Considering the same Figure, we can notice that market-based collaborations are more effective in the process innovation case, since its odds-ratio is equal to 4,67, followed by product and service innovation (3,54) and organizational innovation (2,5). Moreover, activities related to process, product and service innovation are the most

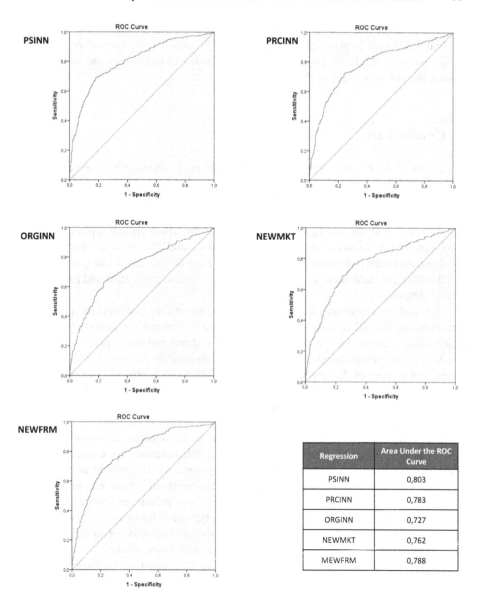

Fig. 3. ROC curves

influenced by the establishment of market-based collaborations. Organizational innovation is still influenced by this type of collaboration, but the likelihood that the innovation could be implemented successfully is slightly lower.

Considering briefly the control variables, we can see that the variable referred to the presence on an internal research and development unit is highly significant and has a strong impact on each of the dependent variables. This is in line with the results of Hung and Chou (2013), where it has been showed that internal R&D units have a moderating effect between OI strategies and innovation performances. The use of an external R&D unit is significant at the 10% level, only in the case of product and service innovation. The variable referred to the number of employees and the financial income are not significant in this model.

6 Conclusions

The results of the research show that only hypothesis H2 should not be rejected, since institutions that are involved in market-based collaborations are more likely to be innovative. Moreover, we can answer to the research question stating that collaborations with users or with suppliers are the best ways to make OI strategies effective. However, the analysis underlines other elements, like which are the types of innovations that Italian financial institutions are pursuing and the effect of market-based collaboration on each of them. More in detail, we have demonstrated that using market-based collaborations is more suitable in order to get process innovation and product and service innovation.

This analysis contributes to the academic articles on financial services innovation, demonstrating that OI framework could be a useful tool in order to analyze innovation patterns that involves both financial institutions and not-traditional players, since we can notice that market-based collaborations are significantly correlated with the innovation performance of the respondents. However, this study represents an initial step in the analysis of OI strategies of institutions, the next step is to understand why this type of collaboration is more effective than others, analyzing in depth different types of market-based cooperation.

Moreover, the analysis could help institutions in the implementation of OI strategies: even if we have showed that marked-based collaborations are overall the most effective, we have discovered that they do not influence all the types of innovation in the same way. Moreover institutions' decision makers should chose this type of collaboration only on the case they want to introduce a new product or services, or if they want to innovate the institution's processes. This type of collaboration is not useful in the case institutions want to implement organizational innovation or in the case they want to introduce a new product or service that is new to the market.

Further studies should consider other measure of innovation performances. The ones presented here are related to the introduction of an innovation, but we do not know if that innovation has brought benefits to the organization, in the process innovation case, or if it is profitable, in the product or service innovation case.

References

Almirall, E., Masanell, R.: Open vs. closed innovation: a model of discovery and divergence. Acad. Manage. Rev. 35(1), 24–47 (2010)

Barras, R.: Interactive innovation in financial and business services: the vanguard of the service revolution. Res. Policy 19(3), 215–237 (1990)

Belderbos R, Gilsing V, Lokshin B, Caree M., Sastre, J.F. The antecedents of new R&D collaborations with different partner types: on the dynamics of past R&D collaboration and innovative performance. Working Paper (2017)

Boot, A.W.A., Thakor, A.V.: Banking scope and financial innovation. Rev. Financ. Stud. **10**(4), 1099–1131 (1997)

Chatenier, E., et al.: Identification of competencies for professionals in open innovation teams. R&D Manage. **40**(3), 271–280 (2010)

Cheng, C.C.J., Shiu, E.C.: The inconvenient truth of the relationship between open innovation activities and innovation performance. Manage. Decis. **53**(3), 625–647 (2014)

Chesbrough, H., Vanhaverbeke, W., West, J.: Open innovation: Researching a New Paradigm. Oxford University Press, Oxford (2006)

Das, P., Verburg, R., Verbraeck, A., Bonebakker, L.: Barriers to innovation within large financial services firms. an in-depth study into disruptive and radical innovation projects at a bank. Eur. J. Innov. Manage. **21**(1), 96–112 (2018)

Frame, W.S., White, L.J.: Empirical studies of financial innovation: lots of talk, little action? J. Econ. Lit. **42**(1), 116–144 (2004)

Garriga, H., Von Krogh, G., Spaeth, S.: How constraints and knowledge impact open innovation. Strateg. Manag. J. **34**, 1134–1144 (2013)

Gomber, P., Kauffman, R.J., Parker, C., Weber, B.W.: On the fintech revolution: interpreting the forces of innovation, disruption, and transformation in financial services. J. Manage. Inf. Syst. **35**(1), 220–265 (2018)

Hung, H.P., Chou, C.: The impact of open innovation on firm performance: the moderating effect of internal R&D and environmental turbulence. Technovation **33**, 368–380 (2013)

Kauppila, O.: Alliance management capability and firm performance: using resource-based theory to look inside the process black box. Long Range Plan. **48**(3), 151–167 (2015)

Khraisha, T., Arthur, K.: Can we have a general theory of financial innovation process? a conceptual review. Financ. Innov. **4**, 4 (2018)

Laursen, K., Salter, A.: Open for innovation: the role of openness in explaining innovation performance among UK manufacturing firms. Strateg. Manage. J. **27**(2), 131–150 (2006)

Lerner, J., Tufano, P.: The consequences of financial innovation: a counterfactual research agenda. Ann. Rev. Financ. Econ. **3**, 41–85 (2011)

Mention, A.: Co-operation and co-opetition as open innovation practices in the service sector: which influence on innovation novelty? Technovation **31**, 44–53 (2011)

Mention, A.L., Torkkeli, M.: Drivers, processes and consequences of financial innovation: a research agenda. Int. J. Entrepreneurship Innov. Manage. **16**(1–2), 5–29 (2012)

Salampasis, D., Mention, A.L., Torkkeli, M.: Open Innovation and collaboration in the financial services sector: exploring the role of trust. J. Bus. Innov. Res. **8**(5), 466–484 (2014)

Sieg, J.H., Wallin, M.W., Von Krogh, G.: Managerial challenges in open innovation: a study of innovation intermediation in the chemical industry. R&D Manage. **40**(3), 281–291 (2010)

Stanko, M.A., Fisher, G.J., Bogers, M.: Under the wide umbrella of open innovation. J. Prod. Innov. Manage **34**(4), 543–558 (2017)

Sveiby, K.E.: Innovation and the global financial crisis – systemic consequences of incompetence. Int. J. Entrepreneurship Innov. Manage. **16**(1–2), 30–50 (2012)

West, J., Bogers, M.: Leveraging external sources of innovation: a review of research on open innovation. J. Prod. Innov. Manage **31**(4), 814–831 (2014)

Other Documents

Accenture: Financial providers: transforming distribution models for the evolving consumers (2017). https://www.accenture.com/t20170111T041601Z__w__/us-en/_acnmedia/Accenture/next-gen-3/DandM-Global-Research-Study/Accenture-Financial-Services-Global-Distribution-Marketing-Consumer-Study.pdffla=en

Financial Times: Moneyfarm aims to profitability by 2019 (2018). https://www.ft.com/content/55ceb160-62c1-3cf2-a9e0-5b28ff2b7a2f

KPMG: The Pulse of Fintech Q4 2016 Global Analysis of Investment in Fintech (2017). https://assets.kpmg.com/content/dam/kpmg/xx/pdf/2017/02/pulse-of-fintech-q4-2016.pdf

Raconteur: Financial services technology (2018). https://www.raconteur.net/financial-services-technology-2018

Identification of Financial Statement Fraud in Greece by Using Computational Intelligence Techniques

Christianna Chimonaki[1], Stelios Papadakis[2], Konstantinos Vergos[1], and Azar Shahgholian[3(✉)]

[1] University of Portsmouth, Portsmouth, UK
{christianna.chimonaki,
konstantinos.vergos}@port.ac.uk
[2] Technology Educational Institute of Crete, Heraklion, Greece
spap@staff.teicrete.gr
[3] Liverpool John Moores University, Liverpool, UK
a.shahgholian@ljmu.ac.uk

Abstract. The consequences of financial fraud are an issue with far-reaching for investors, lenders, regulators, corporate sectors and consumers. The range of development of new technologies such as cloud and mobile computing in recent years has compounded the problem. Manual detection which is a traditional method is not only inaccurate, expensive and time-consuming but also they are impractical for the management of big data. Auditors, financial institutions and regulators have tried to automated processes using statistical and computational methods. This paper presents comprehensive research in financial statement fraud detection by using machine learning techniques with a particular focus on computational intelligence (CI) techniques. We have collected a sample of 2469 observations since 2002 to 2015. Research gap was identified as none of the existing researchers address the association between financial statement fraud and CI-based detection algorithms and their performance, as reported in the literature. Also, the innovation of this research is that the selection of data sample is aimed to create models which will be capable of detecting the falsification in financial statements.

Keywords: Financial statement fraud · Machine learning techniques · Classification

1 Introduction - Background

The stock and bond markets are critical components of a capitalist economy. The efficiency, liquidity, and resiliency of these markets depend on the ability of investors, lenders and regulators to assess the financial performance of businesses that raise capital. Financial statements prepared by such organizations play a very important role in keeping capital markets efficient. They provide meaningful disclosures of where a company has been; where it is currently and where it is going. Most financial statements are prepared with integrity and present a fair representation of the financial

© Springer Nature Switzerland AG 2019
N. Mehandjiev and B. Saadouni (Eds.): FinanceCom 2018, LNBIP 345, pp. 39–51, 2019.
https://doi.org/10.1007/978-3-030-19037-8_3

position of the organization issuing them. These financial statements are based on generally accepted accounting principles (GAAP), which guide the accounting for transactions.

Unfortunately, financial statements are sometimes prepared in ways that intentionally misstate the financial position and performance of an organization. Such misstatements can result from manipulating, falsifying, or altering accounting records. Misleading financial statements cause serious problems in the market and the economy. They often result in large losses for investors, lack of trust in the market and accounting systems, and litigation and embarrassment for individuals and organizations associated with financial statement fraud.

Specifically, according to Wells (2005), financial statement fraud is harmful in many ways. These cases are: Firstly, undermines the reliability, quality, transparency and integrity of the financial reporting process, secondly jeopardizes the integrity and objectivity of the auditing profession, especially auditors and auditing firms for example Andersen, thirdly, diminishes the confidence of the capital markets, as well as market participants, in the reliability of financial information, fourthly makes the capital markets less efficient, fifth adversely affects the nation's economic growth and prosperity, sixth results in huge litigation costs, seventh destroy careers of individuals involved in financial statement fraud, eighth causes bankruptcy or substantial economic losses by the company engaged in financial statement fraud, ninth encourages regulatory intervention, tenth causes devastation in the normal operations and performance of alleged companies, eleventh raises serious doubt the efficacy of financial statement audits and finally erodes public confidence and trust in the accounting and auditing profession.

According to the Association of Certified Fraud Examiners' (ACFE's) in its report to the nation on occupational fraud and abuse (2014), the average financial statement fraud by survey respondents is over US $1 million. Financial statement frauds, such as the WorldCom and Enron frauds, can overstate income by billions of US dollars.

Furthermore "public statistics on the possible cost of financial statement fraud are only educated estimates, primarily because it is impossible to determine actual costs since not all fraud is detected, not all detected fraud is reported, and not all reported fraud is legally pursed" (Rezaee 2002). Therefore, financial statement fraud combined with audit failure, increase the interest of investors, lenders and regulators.

As a result, there is the requirement of investors, lenders and regulators to learn how to detect financial statement fraud more effectively. Therefore, this research aims to investigate how the investors, lenders and regulators can detect financial statement fraud. Section 2 refers in details the specific efforts of previous researchers in detecting financial fraud. Section 3 refers to the proposed methodology. Section 4 is a discussion of our findings. Section 5 provides a conclusion to our research

We employ well-established machine learning techniques to identify the factors which are actually connected with the financial statement fraud. Moreover, we provide intelligent, non-parametric models for the identification of financial fraud observational financial data of any company. Also, this research compares the effectiveness of different tools to detect fraud and find out the gaps existed between the judgments of the experts and different prediction model.

2 Review of Related Literature

There are many different types of fraud, as well as a variety of data mining, and research is continually being undertaken to find the best approach for each case (West 2015). Data mining refers to any method that processes large quantities of data to derive an underlying meaning. Within this classification (West 2015) will consider two categories of data mining: statistical and computational. The statistical techniques are based on traditional mathematical methods, such as logistic regression and Bayesian theory. Computational methods are those who use modern intelligence techniques, such as neural networks and support vector machines. Also (West 2015) consider that these categories share many similarities, but the main difference between them is that computational methods are capable of learning from and adapting to the problem domain, while statistical methods are more rigid. In this research, we examine both types of data mining. Specifically, in this research, we compare the performance of two data mining methods including Naves Bayes, and K-nearest neighbours.

The first researchers (Zhang et al. 1998,) who investigated the fraud detection focused heavily on statistical models such as logistic regression and neural networks. Recent fraud detection research has been far more varied in methods studied, although the former techniques are still popular (West 2015). The most recent studies like Kirkos et al. (2007), Ravisankar et al. (2011), which have examine the financial statement fraud used classification methods to detect fraud. Classification is a data mining method that separates a list of unknown samples into one of several discrete classes (Ngai et al. 2011). Binary classification is a simplified case in which there exists only two possible categories (such as fraudulent and non-fraudulent). In contrast, regression is a traditional statistical method that has been used extensively in data mining for many years. It aims to expose relationships between a dependent variable and a set of independent variables (Ngai et al. 2011).

Kirkos et al. 2007 compared statistical methods with neural networks to identify fraudulent Greek manufacturing companies. Also in 2011, Ravisankar et al. 2011 compared a large range of methods to identify financial statement fraud within Chinese companies. In addition, to supporting vector machines they looked at genetic programming, logistic regression, group method of data handling, and variety of neural networks Ravisankar et al. Also Bose and Wang (2007) compared neural network and decision tree to explore financial statement fraud with financial items from a selection of public Chinese companies. Furthermore, Humpherys et al. (2011) used text mining techniques to investigate the financial statement fraud with managerial statements for US companies. Zhou and Kapoor (2011) looked at common behaviours that are frequently present for financial statement fraud and created a framework to be used for designing detection methods.

The identification of financial fraud is difficult or even impossible by using first principles approach. According to the Institute of Internal Auditors (2001) a fraud examiner commonly uses the following techniques to identify the relationships among the financial data that do not appear reasonable:

- Comparison of current period information with similar information from prior periods. Prior period amounts normally are assumed to be the expectation for the current period. A modification of this comparison is the incremental approach whereby prior period numbers are adjusted for known changes, such as significant purchases or sales of assets and changes in lines and volumes of business.
- Comparison of current period information with budgets or forecasts. This comparison should include adjustments for expected unusual transactions and events.
- Study of relationships among elements of information. Certain accounts vary in relation to others, both within a financial statement and across financial statements. For instance, commissions are expected to vary directly in relation to sales.
- Study of relationships of financial information with the appropriate non-financial information. Non-financial measures are normally generated from an outside source. An example would be retail stores where sales are expected to vary with the number of square feet of shelf space.
- Comparison of information with similar information from the industry in which the organization operates. Industry averages are reliable in stable industries. Unfortunately, industry trade associations require months to compile, analyze, and publish information; therefore, the data may not be timely.
- Comparison of information with similar information from other organizational units. A company with several stores might compare one store with another store. The "model" store should be sufficiently audited to assure that it is an appropriate standard.

As we can conclude for the above procedure about the techniques which a fraud examiner uses to detect financial fraud appear many gaps. On the other hand computational intelligence and statistics help to anticipate and quickly detect fraud and take immediate action to minimize costs.

However, we assume that there exists a relationship between specific financial attributes and the existence or absence of financial fraud (outcome). This potential relationship between these factors and the outcome is not exactly known due to the inherent uncertainty Parsons (1996), Ren et al. (2009), of the financial data. As a consequence, we are dealing with the problem as a 'black box' system. The input of the system is a set of specific attributes (factors), while its output is the outcome of these attributes, caused by the system in a way which is not exactly known. The only knowledge we have about the operation of the system arises from specific observations regarding what outcome causes specific inputs (attributes). The target of modeling is building a model (i.e. a mathematical function) for simulating the unknown system. That is a model that delivers the same outcome as the unknown system on the given data set of observations.

Over the years, various computational methods have been used for fraud detection and, like other similar problems; successful implementation of the detection methods depends on having a clear understanding of the problem domain. While some prior researchers have focused on the common issues such as problem representation for machine learning techniques problems, in general, there has been almost no analysis from the perspective of fraud detection which we aim to address here. The implementation of these techniques follows the same information flow of machine learning techniques processes in general.

3   The Proposed Methodology

We formulate the problem of financial fraud detection as a classification problem, assuming that the existence or the absence of financial fraud depends on specific quantitative financial attributes. These attributes, listed in Table 1, are the input to the classifier. The output of the classifier is either '1' = FFSs (Financial Fraud Statement) or '0' = Non-FFSs, indicating the existence or the absence of fraud, respectively. If sufficient historical data (instances, in the form attribute-label) exist, then the classifier's workflow can be directed at increasing the chances to capture the opportunities for preventing loss by identifying and verifying potential financial fraud.

Table 1. The number of firms per sector

Sector	Number of firms/sector
Industrial Goods & Services	19
Retail	13
Construction & materials	33
Media	14
Oil & Gas	3
Personal & Household Goods	47
Travel & Leisure	12
Technology	27
Telecommunications	1
Food & Beverage	28
Health Care	8
Chemicals	9
Basic Resources	17
Total	**231**

In this research, we follow CRISP-DM approach which follows the following steps: (i) Business Understanding, (ii) Data Understanding, (iii) Data preprocessing, (iv) Modeling, (v) evaluation, and (vi) Deployment. Business understanding phase was presented in Sect. 2. In this section, data collection, data understanding and modeling are discussed. Section 4; explain the findings, evaluation and deployment phases.

3.1   Data Collection/Description

A sufficient number of samples should be collected after the definition of candidate attributes. These samples are raw data and usually needs preprocessing for detecting potential outliers and missing values. Another important preprocessing data step is the normalization of attributes.

The selection of data sample is aimed to create models which will be capable of detecting the falsification in financial statements. For this reason, several factors have been examined. One of the most important factors is the sector of enterprises because the sector of enterprises affects their financial profile. Our main sources for data were the published financial statements and their notes from the Athens Stock Exchange database.

Initially, our sample contained data from 231 Greek listed on the Athens Stock exchange since 2002–2015. Our sample contains 2469 observations. We analyze the number of firms per sector in Table 1, after excluding the sectors of banking, utilities, and financial services, from the sample.

According to Spathis et al. (2002b), and Kirkos et al. (2007), the classification of the financial statement as fraud was based on the following parameters:

- The inclusion in the auditors' reports of opinions of serious doubt as to the correctness of accounts,
- The observations by the tax authorities regarding serious taxation intransigencies which seriously alter the company's financial statements,
- The application of Greek legislation regarding negative net worth,
- The inclusion of the company in the Athens Stock Exchange categories of "under observation" and "negotiation suspended" for reasons associated with falsification of the company's financial data and
- The size of the auditor firm.

After the selection of the fraud sample, we searched for a non-fraud sample from the same sources. The choice of the non-fraud enterprises was carried out by using the matching method Hunt and Ord (1988), Sibley and Burch (1979). The matching method is a common practice in financial classification researches such as bankruptcy, mergers, acquisitions, etc. Beaver (1966),. There are two main reasons which we use the method of matching. The first reason is the high cost and the time which is needed for the selection of sample Bartley and Boardman (1990) and the second reason is the higher information which contained in this sample in compare of a random sample Cosslett (1981), and Palepu (1986).

Therefore, the main criterion for the similarity of the two samples is the period Stevens (1973). The criterion of period refers to the changes in a country's macroeconomic environment and has an impact on economic conditions and business decision making. Also, there is one more main criterion which is the sector and the total assets. Stice (1991) referred that the sector and the size are the most important factors for the matching method.

On the other hand, the matching method has accepted criticisms. Ohlson (1980) refers that the criteria which used for the matching method tend to be arbitrary. Also, Ohlson (1980) refers that there is not absolutely clear the advantages process of the matching method. Ohlson (1980) suggests that is more preferable to use the different factors as independent variables of the sample than to use for the purpose of matching method.

3.2 Candidate Attributes

This paper adopted the related attributes based on prior researchers, who study the FFS. Such work carried out by Spathis et al. (2002a, b), Fanning and Cogger (1998), Persons (1995), Stice (1991), Feroz et al. (1991), Loebbecke et al. (1989), and Kinney and

McDaniel (1989) contained suggested indicators of FFS. So there are a number of attributes which considered more possible to lead in the falsification of the financial statement. The financial ratios, examined in this research appear in Table 2.

Table 2. The list and description of candidate attributes.

Attribute	Description	
x_1	Total Debt	Solvency ratios
x_2	The logarithm of Total Debt	Solvency ratios
x_3	Equity	Structure ratios
x_4	Debt to Equity	Solvency ratios
x_5	Total Debt/Total Assets	Solvency ratios
x_6	Long Term Debt/Total Assets	Solvency ratios
x_7	Short-Term Debt/Total Assets	Solvency ratios
x_8	Account Receivable/Sales	Activity ratios
x_9	Inventory/Sales	Activity ratios
x_{10}	Inventory/Total Assets	Activity ratios
x_{11}	Sales Growth	Activity ratios
x_{12}	Sales	Activity ratios
x_{13}	Gross margin	Profitability ratios (Return on sales)
x_{14}	Sales minus Gross Margin	Activity ratios
x_{15}	Total assets	Structure ratios
x_{16}	The logarithm of Total Assets	Structure ratios
x_{17}	Net fixed assets/total assets	Structure ratios
x_{18}	Gross Profit/Total Assets	Profitability ratios (Return on investment)
x_{19}	Net Profit/Total Assets	Profitability ratios (Return on investment)
x_{20}	Net Profit/Sales	Profitability ratios (Return on sales)
x_{21}	Working Capital	Liquidity ratios
x_{22}	Working Capital/Total Assets	Liquidity ratios
x_{23}	Sales to total assets	Activity ratios
x_{24}	Current Assets/Current Liabilities	Liquidity ratios
x_{25}	Net Income/Fixed Assets	Profitability ratios (Return on investment)
x_{26}	Cash/Total Assets	Liquidity ratios
x_{27}	Quick Assets/Current Liabilities	Liquidity ratios
x_{28}	Earnings Before Interest and Taxes	Profitability ratios (Return on sales)
x_{29}	Ebit/Total Assets	Profitability ratios (Return on investment)
x_{30}	Equity/Total Liabilities	Structure ratios
x_{31}	Z-score	Profitability ratios (Return on investment)
x_{32}	Inventory	Activity ratios
x_{33}	Net profit after tax	Profitability ratios (Return on sales)
x_{34}	Sector	
x_{35}	P/E	
x_{36}	Price/book value	Investment ratios

3.3   Data Preprocessing

Data preprocessing involves several steps, for preparing cleansing and normalizing the raw data before being used for modeling. Missing values is one of the most common issues that the data preprocessing should face. In this work, we entirely remove a sample from a data set if one or more attributes of the sample have missing values. In addition, in this work, we performed the normalization step by linearly mapping each attribute's value from its actual range within the interval $[0, 1]$. In the next step, we considered as outliers those instances (companies) having extreme or out of feasible range values for some attributes. Outliers were removed from the data set before applying any modeling technique.

We use wrapper based methods as they tend to deliver more accurate results than filter based ones Monroe and The (1993). A particular model is used as wrapper and different subsets of attributes are sequentially presented to it according to forward inclusion approach.

3.4   Description of Employed Models (Wrappers)

We use particular models from established paradigms of machine learning and from statistics. More specifically, we use K-Nearest Neighbor as a representative from "instance-based learning"; From statistics, we use Naïve Bayes method from the "Bayesian paradigm". Although, a lot of variations of each model exist, however, we apply the "principal" model which we consider as "representative" of each paradigm.

The main advantage of K-Nearest Neighbor Classifier is a very simple classifier that works well on basic recognition problems. The main disadvantage of this approach is that the algorithm must compute the distance and sort all the training data at each prediction, which can be slow if there are a large number of training examples. On the other hand, the first advantage of Naïve Bayes Classifier is fast to train fast to classify, Second in not sensitive to irrelevant features. Thirdly it handles real and discrete data and finally the Naïve Bayes Classifier handles streaming data well. Also the main disadvantage is that it assumes strong feature independence assumption.

4   Experimental Results

4.1   Comparison with Factor Importance

Overall, in Table 3 appears the comparison results from all the methods of machine learning techniques. Also, in Table 3 shows the fraud factors in different methods and the comparison of empirical data result. In addition, Table 3 indicates the importance of attributes included in prediction models. The most important category of fraud detection is "poor performance". All factors effects are consistent with prior researches. The top seven fraud factors are a log of Total Debt, Equity, Debt to Equity, a log of total assets, net fixed assets to total assets, cash to total assets and sector. Furthermore, the Profitability, Liquidity, Solvency, Activity and Structure ratios are significant predictors for fraud detection. Specifically, the significant ratios which are the most important for fraudulent financial statements appeared in Table 3 and analyzed following.

Leverage proxies is a significant result as an indicator for fraud analysis. These ratios are consistent with Spathis et al. (2002b) while and Fanning and Cogger (1998) which suggest that firms with higher debt to equity ratios would be a good indicator for fraudulent firms. Furthermore, it means that firms with a high total debt to total equity value have an increased probability to be classified as fraudulent firms. Previous studies such as Persons (1995) supported that the high debt structure it is possible to motivate in the FFS. In addition, Loebbecke et al. (1989) concluded in their research that 19% of firms of their sample appeared solvency problems.

Lower liquidity may be an incentive for managers to engage in fraudulent financial statements. This argument is supported by Kreutzfeldt and Wallace (1986) who discovered that firms with liquidity problems have significant more errors in their financial statement than firms without liquidity problems. In this research, the most important liquidity ratios which associate with the fraudulent financial statement are the Working capital, Current assets to Current liabilities and Cash to Total Assets.

Table 3. Comparison with factor & predict the importance

Attributes	Result of KNN	Result of NB	Total
Total Debt	X		1
The logarithm of Total Debt		X	1
Equity	X		1
Debt to Equity	X		1
Total Debt/Total Assets		X	1
Long Term Debt/Total Assets		X	1
Short-Term Debt/Total Assets		X	1
Account Receivable/Sales			0
Inventory/Sales			0
Inventory/Total Assets			0
Sales Growth	X		1
Sales	X		1
Gross margin			0
Sales minus Gross Margin	X		1
Total assets		X	1
The logarithm of Total Assets			0
Net fixed assets/total assets		X	1
Gross Profit/Total Assets		X	1
Net Profit/Total Assets			0
Net Profit/Sales			0
Working Capital	X		1
Working Capital/Total Assets			0
Sales to total assets			0
Current Assets/Current Liabilities		X	1
Net Income/Fixed Assets	X		1

(continued)

Table 3. (*continued*)

Attributes	Result of KNN	Result of NB	Total
Cash/Total Assets		X	1
Quick Assets/Current Liabilities			0
Earnings Before Interest and Taxes			0
Ebit/Total Assets		X	1
Equity/Total Liabilities		X	1
Z-score			0
Inventory	X		1
Net profit after tax			0
Sector	X	X	2
P/E			0
Price/Book Value			0
CVSR -ACCURANCY	**89,11**	**68,29**	

Furthermore, lower profit may give management incentive to overstate revenue or understate expenses. Kreutzfeldt and Wallace (1986) discovered that firms with profitability problems have significantly more errors in the financial statement than firm without profitability problems. This approach is based on the expectation that management will be able to maintain or improve past levels of profitability Summers & Sweeney (1998). If this expectation is not met by actual performance, then it motivates the fraudulent financial statement. Financial distress is a motivation for fraudulent financial statements Loebbecke et al. (1989), Kreutzfeldt and Wallace (1986). In this research, the most important profitability ratios for FFS are gross profit to total assets, net profit to total assets, net income to fixed assets and EBIT to total assets.

Capital Turnover proxies by receivables to revenue also have significant results. High ratios of account receivables to sales and inventory to sales are consistent with research suggesting that accounts receivables are an asset with a higher incidence of manipulation. Also, asset composition proxies by inventory to total assets indicate significant results. In addition, our research concludes that the size of the firm is statistically significant and measured by total assets. Finally, ratios sales growth, sales to total assets sales minus gross margin inventory net fixed assets to total assets equity to total liabilities, and P/E are significant in the detection of the fraudulent financial statement.

This result supported by the result of the research with the rate of correct classification which analyzed in the next section.

4.2 Comparison with Predict Performance

Performance evaluation is the final step of the framework which is used for measuring the performance and judging the efficacy of machine learning techniques.

The pre-processed dataset was further randomly divided into training and testing sets via K-fold cross-validation. A typical experiment uses K = 5. The sample was divided 5-fold via stratified 5-fold cross-validation. Each fold contained equal numbers of fraud and non-fraud cases. Each fold of the sample was used individually to define parameters and train classifiers, while the remaining five folds were used as test sets to assess the sample performance. After the parameters were set and the classifiers have trained the methods were evaluated by applying them to the test sets. Finally, the average classification accuracy of the test sets was calculated. After preparation of the 5-fold cross validation datasets these datasets were used by the two classifiers. The proposed ensemble of classifiers was developed and validated based on the classifier results.

Besides classification accuracy, this research also used misclassification cost. Generally, misclassification cost is associated with two error types. A type error I occur when a non-fraud case is classified as a fraud class. Meanwhile, a type II error is committed when a fraud case is classified as a non-fraud class. The misclassification costs associated with type II errors are reportedly much higher than those associated with type I error West et al. (2014). Classifying a fraud case into a non-fraud class may result in incorrect decisions about economic damage. Moreover, classifying a non-fraud case into a fraud class may result in expenses and excess time associated with the additional investigation.

The 5-fold cross validation performances of the six classification methods were calculated and compared. The KNN has the higher average accuracy (89,11%), and Naives Bayes has the lowest accuracy (68,29%) respectively.

The confusion matrix for KNN and NB are presented in Tables 4 and 5. Also, performance matrix indicating the sensitivity (type I error) and specificity (type II error) of the two methods which are used in this research. Sensitivity (type I error) and specificity (type II error) have been used as a metrics for performance evaluation. The sensitivity is the measure of the proportion of the number of fraudulent companies predicted correctly as fraudulent by a particular model to the total number of actual fraudulent companies. The specificity is the measure of the proportion of the number of non-fraudulent companies predicted as non-fraudulent by a model to the total number of actual non-fraudulent companies. In both cases, we presented the average accuracies, Sensitivity, specificity, accuracy, error rate, precision (Table 6).

Table 4. Confusion matrix for KNN

Observed	Predicted-classified as KNN	
	FFSs	Non-FFSs
FFSs	1300 (correct classification)	113 (type II error)
Non- FFSs	132 (type I error)	924 (correct classification)

Table 5. Confusion matrix for Naive Bayes

Observed	Predicted-classified as Naive Bayes	
	FFSs	Non-FFSs
FFSs	1299 (correct classification)	114 (type II error)
Non- FFSs	669 (type I error)	387 (correct classification)

Table 6. Sensitivity, specificity, accuracy, error rate, precision

Sensitivity	Specificity	CVSR -accuracy	Misclassification rate- error rate	Precision
K - Nearest Neighbours (KNN)				
87.50%	92.00%	89,11%	9.92%	89.10%
Naive Bayes (NB)				
36.65%	91.93%	68,29%	31.71%	77.25%

5 Conclusion

Reasons for committing financial statement fraud include improving stock performance, reducing tax obligations or as an attempt to exaggerate performance due to managerial pressure Ravisankar, et al. (2011). Financial statement fraud can be difficult to diagnose because of a general lack of understanding of the field, the infrequency in which it occurs, and the fact that it is usually committed by knowledgeable people within the industry who are capable of masking their deceit Maes et al. (2002). This research studied intelligent approaches to fraud detection, both statistical and computational. There is also the opportunity to examine the performance of existing methods by adjusting their parameters, as well as the potential to study cost-benefit analysis of computational fraud detection. Finally, further research into the differences between each type of financial fraud could lead to a general framework which would greatly improve the accuracy of intelligent detection methods.

References

Association of Certified Fraud Examiners' (ACFE's) in its report to the nation on occupational fraud and abuse (2014)

Bartley, J.W., Boardman, C.M.: The relevance of inflation adjusted accounting data to the prediction of corporate takeovers. J. Bus. Financ. Acc. **17**(1), 53–72 (1990)

Beaver, W.H.: Financial ratios as predictors of failure. J. Acc. Res. **4**, 71–111 (1966)

Bose, I., Wang, J.: Data mining for detection of financial statement fraud in Chinese Companies. In: Paper Presented at the International Conference on Electronic Commerce, Administration, Society and Education, Hong Kong, pp. 15–17, August 2007

Cosslett, S.R.: Efficient estimation of discrete-choice models. In: Structural Analysis of Discrete Data with Econometric Applications (1981)

Fanning, K.M., Cogger, K.O.: Neural network detection of management fraud using published financial data. Int. J. Intell. Syst. Acc. Financ. Manag. **7**(1), 21–41 (1998)

Feroz, E.H., Park, K., Pastena, V.S.: The financial and market effects of the SEC's accounting and auditing enforcement releases. J. Acc. Res. **29**, 107–142 (1991)

Humpherys, S.L., Moffitt, K.C., Burns, M.B., Burgoon, J.K., Felix, W.F.: Identification of fraudulent financial statements using linguistic credibility analysis. Decis. Support Syst. **50**, 585–594 (2011)

Hunt, H.G., Ord, J.K.: Matched pairs discrimination: methodology and an investigation of corporate accounting policies. Decis. Sci. **19**(2), 373–382 (1988)

Kinney, W.R., McDaniel, L.S.: Characteristics of firms correcting previously reported quarterly earnings. J. Acc. Econ. **11**(1), 71–93 (1989)

Kirkos, E., Spathis, C., Manolopoulos, Y.: Data mining techniques for the detection of fraudulent financial statements. Expert Syst. Appl. **32**(4), 995–1003 (2007)

Kreutzfeldt, R., Wallace, W.: Error characteristics in audit populations: their profile and relationship to environment factors. Auditing J. Pract. Theory **6**, 20–43 (1986)

Loebbecke, J.K., Eining, M.M., Willingham, J.J.: Auditors experience with material irregularities-frequency, nature, and detectability. Audit. J. Pract. Theory **9**(1), 1–28 (1989)

Maes, S., Tuyls, K., Vanschoenwinkel, B., Manderick, B.: Credit card fraud detection using Bayesian and neural networks. In: Proceedings of the 1st International Naiso Congress on Neuro Fuzzy Technologies (2002)

Ngai, E., Hu, Y., Wong, Y., Chen, Y., Sun, X.: The application of data mining techniques in financial fraud detection: a classification framework and an academic review of literature. Decis. Support Syst. **50**, 559–569 (2011)

Ohlson, J.A.: Financial ratios and the probabilistic prediction of bankruptcy. J. Acc. Res. **18**, 109–131 (1980)

Palepu, K.G.: Predicting takeover targets: a methodological and empirical analysis. J. Account. Econ. **8**(1), 3–35 (1986)

Parsons, S.: Current approaches to handling imperfect information in data and knowledge bases. IEEE Trans. Knowl. Data Eng. **8**(3), 353–372 (1996)

Persons, O.S.: Using financial statement data to identify factors associated with fraudulent financial reporting. J. Appl. Bus. Res. **11**(3), 38 (1995)

Ravisankar, P., Ravi, V., Rao, G.R., Bose, I.: Detection of financial statement fraud and feature selection using data mining techniques. Decis. Support Syst. **50**(2), 491–500 (2011)

Ren, J., Lee, S.D., Chen, X., Kao, B., Cheng, R., Cheung, D.: Naive bayes classification of uncertain data. In: 2009 Ninth IEEE International Conference on Data Mining, 2009, pp. 944–949 (2009)

Rezaee, Z.: In Financial statement fraud: prevention and detection. Wiley, Hoboken (2002)

Sibley, A., Burch, E.E.: Optimal selection of matched pairs from large data bases. Decis. Sci. **10**(1), 62–70 (1979)

Spathis, C., et al.: Detecting false financial statements using published data: some evidence from Greece. Manag. Auditing J. **17**(4), 179-191 (2002a)

Spathis, C., Doumpos, M., Zopounidis, C.: Detecting falsified financial statements: a comparative study using multicriteria analysis and multivariate statistical techniques. Eur. Acc. Rev., **11**(3), 509–535 (2002b)

Stevens, D.L.: Financial characteristics of merged firms: a multivariate analysis. J. Financ. Quant. Anal. **8**(02), 149–158 (1973)

Stice, J.D.: Using financial and market information to identify pre-engagement factors associated with lawsuits against auditors. Acc. Rev. **66**, 516–533 (1991)

The Institute of Internal Auditors: Practice Advisory 2320-1. Analysis and Evaluation, January 5 2001 (2001)

Wells, J.T.: Principles of Fraud Examination. Wiley, Hoboken (2005)

West, J., Bhattacharya, M., Islam, R.: Intelligent financial fraud detection practices: an investigation. In: Proceedings of the 10th International Conference on Security and Privacy in Communication Networks (SecureComm 2014) (2014)

West, J., Bhattacharya, M., Islam, R.: Intelligent financial fraud detection practices: an investigation. In: International Conference on Security and Privacy in Communication Systems, pp. 186–203. Springer, Cham (2015)

Zhang, G., Eddy Patuwo, B., Hu, M.V.: Forecasting with artificial neural networks: the state of the art. Int. J. Forecast. **14**, 35–62 (1998)

Zhou, W., Kapoor, G.: Detecting evolutionary financial statement fraud. Decis. Support Syst. **50**, 570–575 (2011)

What Sort of Asset? Bitcoin Analysed

Shaen Corbet[1], Brian Lucey[2], Maurice Peat[3(✉)], and Samuel Vigne[4]

[1] DCU Business School, Dublin City University, Dublin 9, Ireland
[2] Trinity Business School, Trinity College Dublin, Dublin 2, Ireland
[3] University of Sydney Business School, Sydney, NSW, Australia
`maurice.peat@sydney.edu.au`
[4] Queen's Management School, Queen's University Belfast,
Belfast BT9 5EE, Northern Ireland

Abstract. Early analysis of Bitcoin concluded that it did not meet the economic conditions to be classified as a currency. Since this analysis interest in bitcoin has increased substantially. We investigate whether the introduction of futures trading in bitcoin is able to resolve the issues that stopped bitcoin from being considered a currency. Our analysis shows that spot volatility has increased following the announcement of the futures contracts, the futures contracts are not an effective hedging instrument and that price discovery is driven by uninformed investors in the spot market. The conclusion that bitcoin is a speculative asset rather than a currency is not altered by the introduction of futures trading.

Keywords: Cryptocurrencies · Futures markets · Volatility ·
Speculative assets · Currencies

1 Introduction

An early analysis of bitcoin by [22] concluded that it was not a currency, but rather a speculative asset. In drawing this conclusion Yermak argued that bitcoin failed to satisfy all three of the functions of money: a medium of exchange, unit of account and store of value. The finding that bitcoin has no intrinsic value, in work that found speculative bubbles in bitcoin prices, by [4] supports this conclusion. A recent innovation in the bitcoin trading environment is the introduction of futures contracts by the Chicago Mercantile Exchange (CME) and Chicago Board Options Exchange (cboe) in December 2017. The high volatility of bitcoin prices was identified by Yermak as a feature which lead to bitcoin not being a useful unit of account. We explore whether the introduction of futures contracts has materially reduced the volatility of bitcoin prices on the spot market. We also explore which of the markets is more important in setting the bitcoin price and the hedging effectiveness of the new futures contract. A finding that volatility has not fallen, hedging opportunities are limited and that the bitcoin price is set by uninformed investors on the spot market will reinforce Yermak's conclusion that bitcoin is a speculative asset rather than a currency.

© Springer Nature Switzerland AG 2019
N. Mehandjiev and B. Saadouni (Eds.): FinanceCom 2018, LNBIP 345, pp. 52–65, 2019.
https://doi.org/10.1007/978-3-030-19037-8_4

What is a crypto currency? A standard electronic transaction will involve an intermediary such as a bank or third-party company who will ensure that the value of the transaction is transferred between the two parties. In a crypto currency transaction there is no third-party. Each transaction is recorded in a distributed database maintained on a peer-to-peer network. Each transaction in the database records the buyer and seller of the unit of currency. The database is updated so each node of the network maintains a copy of all transactions. Units of currency are created as rewards for parties who provide the required computing infrastructure (known as mining). Existing and new coins are passed from user to user by adding a new transaction to the transaction record. This approach means that it is possible to track each coin from creation through to its current owner. The system uses public private key encryption to confirm transactions and provide anonymity to users of the system; the only identifier recorded with transfers is the public encryption key of each party to the transaction.

Bitcoin was proposed as a peer-to-peer online payment platform, [15]. The messages which contain the information required for a transaction are called coins. The bitcoin network facilitates the creation of coins and a mechanism for recording the transfer of coins between the users of the network. The design of the system uses cryptography and distribution of the underlying transaction data, allowing the transfer of coins to occur without a third-party intermediary. The mechanism used to solve the double spending problem inherent in electronic value transfer systems is a distributed timestamp service. Every transaction (transfer of coins) is broadcast to the network, the servers supporting the network aggregate these transaction into a block, which contains a link to the last block of transactions (thus a block chain) and its hash. A block is verified by generating a hash code (a unique numeric value) that is dependent on the contents of the block. The key is required to satisfy a system generated constraint, generally specified as the number of leading zeros in the hash key. A section of each block has been set aside to hold a number that can be altered, thus altering the content of the block and the hash key of the block. The first node to find a conforming hash code broadcasts the completed block to the network. Completing a block earns a number of new coins for the operator of the network node, a built-in incentive for node operators to participate in the network.

The information in a coin records the transfer of ownership of the coin. These transfers use public private key cryptographic methods to authenticate ownership of coins by the seller and to create a record of the recipient of the coin. A public key is designed to be widely shared, any message that is encrypted with this key can only be decrypted with the associated private key. Anyone can encrypt a message that can only be read by the private key holder. A message encrypted with a private key can only be decrypted with its associated public key, allowing the message to be confirmed as authentic. The coin network uses private key encryption to authenticate that the transfer instruction came from the current owner of the coin. Someone transferring coins publishes a message to the network indicating the number of coins being transferred with a reference to the transaction that transferred these coins to the sender. Part of the message is encrypted with a private key to authenticate the transaction, the public key of the recipient of the transfer is also included in the message. The network identifies the parties to a transaction by their public key, which are effectively account numbers. This is why the network is called pseudo-anonymous, actual names of transactors are not known but knowledge of public keys facilitates the identification of all transactions by the user of that key.

By construction the bitcoin network simply records the transfer of coins and the creation of new coins by network nodes. Other infrastructure is necessary for the network to acquire economic significance, [2] provide an overview of economic and governance issues associated with Bitcoin. As with equities, organised exchanges facilitate the buying and selling of coins for currency or other assets. Exchanges provide an order recording and matching mechanism to put buyers and sellers together. Following the matching of orders the transfer of the bitcoins proceeds by the usual mechanism, the transfer of funds from the buyer to the seller happens through the payments platform used by the exchange. This will typically be provided by a bank, card supplier or alternate payment platform. For vendors to accept bitcoin as payment a gateway between the bitcoin network and standard payment systems which has access to reliable bitcoin prices in a currency acceptable to merchants is required. Risk management services in the form of futures trading will also be provided as external infrastructure by futures exchanges. None of this economic infrastructure is inherent in the design and operation of the base Bitcoin network.

In December 2017 trading in futures contracts on bitcoins commenced on the Chicago Mercantile Exchange (CME) and the Chicago Board Options Exchange (cboe). On December 1 both exchanges announced a bitcoin futures contract. The cboe contract commenced trading on December 10, each contract is for one bitcoin. Three aspects of the introduction of futures on the spot market will be explored. Firstly the impact of futures trading on spot volatility is examined. Secondly the hedging effectiveness of the futures contracts is evaluated. Finally the flow of information between the spot and futures markets is documented.

2   Data

The CME contract commenced trading on December 18, each contract is for 5 bitcoins. Both contracts are cash settled in USD. Shown in Table 1 are stylized facts of these two contracts. Using data sampled at a one minute frequency from the cboe futures contract, sourced from Thomson Reuters Tick History, and bitcoin price data from coinmarketcap.com[1], we will explore the impact of the introduction of risk management tools on the pricing and risk characteristics of the spot bitcoin market. From the 1-minute transaction prices we calculate the log return for each period, $r_t = ln(P_t/P_{t-1})$, which is presented in Fig. 1. This return series is used in the analysis which follows.

---

[1] A website which collects Bitcoin data from multiple exchanges and combines it to form a weighted average.

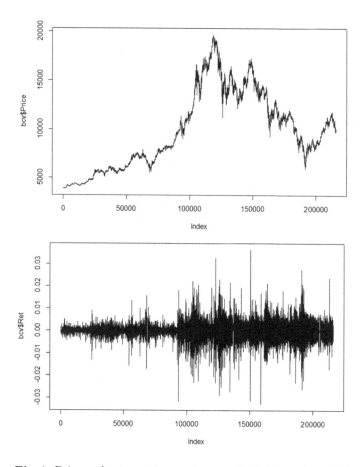

Fig. 1. Price and returns time series over the full sample period

The characteristics of the bitcoin data covering the period from 26 September 2017 to 22 February 2018 can be found in Table 2. Statistics for the full period and for sub-samples before and after the introduction of futures trading are presented.

Both the descriptive statistics and the graphs show that there has been a change in the distributional characteristics of bitcoin returns. The mean has changed sign and the standard deviation has doubled. This change in volatility is evident in the time series plot of the returns. We wish to determine if these changes are significant and if possible to date the change.

3   Analysis

The impact of the introduction of futures trading on volatility in the underlying spot market has been investigated for stocks, foreign exchange, interest rates

Table 1. Stylised facts based on Cboe and CME Bitcoin Futures

Variable	Cboe futures	CME futures
Product code	XBT	BTC
First traded	10th of December 2017	18th of December 2017
Contract unit	1 Bitcoin	5 Bitcoins
Minimum price fluctuation	10.00 points USD/XBT (equal to $10.00 per contract)	$5.00 per bitcoin = $25.00 per contract
Position limits	A person: (i) may not own or control more than 5,000 contracts net long or net short in all XBT futures contract expirations combined and (ii) may not own or control more than 1,000 contracts net long or net short in the expiring XBT futures contract, commencing at the start of trading hours 5 business days prior to the Final Settlement Date of the expiring XBT futures contract	1,000 contracts with a position accountability level of 5,000 contracts
Price limits	XBT futures contracts are not subject to price limits	7% above and below settlement price, +/−13% previous settlement, +/−20% for prior settlement
Settlement	The Final Settlement Value of an expiring XBT futures contract shall be the official auction price for Bitcoin in U.S. dollars determined at 4:00 p.m. Eastern Time on the Final Settlement Date by the Gemini Exchange Auction	Cash settled by reference to Final Settlement Price

and commodities. The empirical evidence is mixed. [8] examined stock market volatility before and after the introduction of stock index futures trading in 25 markets. They found a noticeable increase in volatility in the U.S. and Japan. In the remaining 23 markets there was a negligible effect or the volatility fell. A recent study of the introduction of futures on European real estate indices by [12] found that the volatility of the indices fell after the introduction of the futures contracts. The question is an empirical one.

To test for a change in volatility the change point detection methods that have been developed in the process control literature are applied, [20] and [19]. With a sequence of observations $x_1, x_2, \ldots$ drawn from random variables $x_1, x_2, \ldots$ which undergo a change in distribution at time τ, the observations are distributed:

$$X_i \sim \begin{cases} F_0 & \text{if } i \leq \tau_1 \\ F_1 & \text{if } i > \tau_1 \end{cases} \tag{1}$$

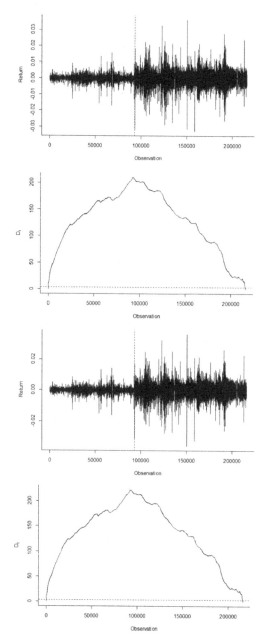

Note: The above figure presents the Raw Returns Mood Statistics (Top Panel) and GARCH(1,1) Residuals Mood Statistic (Bottom Panel) respectively. These two nonparametric statistics represent the Mood statistic for change in volatility (scale) and a Lepage type statistic which tests for a change in location and scale respectively. The implementation of these statistics for change point detection the cpm package was used to establish the existence and location of a change point in the bitcoin price series. Both the Mood and Lepage statistics indicate there is a significant change in the distribution, driven by the increase in volatility. The date of the change is 29 November 2017, two days before the official announcement of the commencement dates for futures trading.

Fig. 2. Change point detection

Table 2. Descriptive statistics for bitcoin prices and returns

Panel A - *Full Sample*	Price	Return
Mean	9,862.048	4.26E-06
Standard error	8.579189	4.33E-06
Median	9,291.53	1.21E-06
Mode	15,000	0.000000
Standard deviation	3,984.44	0.002009
Sample variance	15,875,760	4.04E-06
Kurtosis	−0.89573	11.46425
Skewness	0.39184	−0.08776
Range	15,800.5	0.069144
Minimum	3,865.23	−0.03291
Maximum	19,665.73	0.036236
Count	215,696	215,696
Panel B - *Pre Futures Introduction*	Price	Return
Mean	7,812.788	1.3E-05
Standard error	10.39188	3.96E-06
Median	6,671.42	1.1E-05
Mode	16,500	0.000000
Standard deviation	3,559.035	0.001357
Sample variance	12,666,728	1.84E-06
Kurtosis	0.845098	26.04
Skewness	1.322531	−0.43248
Range	14,152.89	0.053846
Minimum	3,865.23	−0.03166
Maximum	18,018.12	0.022191
Count	117,294	117,294
Panel C - *Post Futures Introduction*	Price	Returns
Mean	12304.74	−6.1E-06
Standard error	9.418187	8.22E-06
Median	11,683.09	0.000000
Mode	15,000	0.000000
Standard deviation	2,954.4	0.00258
Sample variance	8,728,479	6.66E-06
Kurtosis	−0.58609	6.020228
Skewness	0.302747	−0.00718
Range	13,741.01	0.069144
Minimum	5,924.72	−0.03291
Maximum	19,665.73	0.036236
Count	98,402	9,8402

That is the variables have some distribution F_0 before the change point at $t = \tau$ and a different distribution F_1 after. It is assumed that the observations are univariate, real-valued, and independent. Begin by considering the problem of testing for a change in distribution at some given time k. This leads to the following hypothesis test:

$$H_0 : X_i \sim F_0(x; \theta_0), i = 1, 2, ..., n$$
$$H_1 : X_i \sim \begin{matrix} F_0(x; \theta_0), i = 1, 2, ..., k \\ F_1(x; \theta_1), i = k+1, k+2, ..., n \end{matrix} \tag{2}$$

Where the θ_i are the parameters of each distribution, which will generally be unknown. This is a problem which can be solved using a two sample test, the test chosen will depend on what is known about the distribution of the observations. To avoid making distributional assumptions, non-parametric tests can be employed such as the Mann-Whitney test for mean (location) shifts, the [14] test for variance (scale) shifts and the [13] test for more general distributional changes. After choosing a two sample test statistic, $D_{k,n}$, its value is computed by splitting the observations into two samples at observation k. If its value exceeds the chosen critical value then the null of the samples having identical distributions is rejected and we conclude that there is a change point at observation k. When the location of the change point is not known in advance we do not know what value of k to use in the test. In this case the value of the test statistic, $D_{k,n}$, is computed for all values of k, $1 < k < n$. That is, for every possible split of the data into two samples the chosen test is conducted and the maximum value of the standardised test statistics obtained, D_n, is used for inference. Where:

$$D_n = \max_{k=2,...,nn-1} \left| \frac{\tilde{D}_{k,n} - \mu_{\tilde{D}_{k,n}}}{\sigma_{\tilde{D}_{k,n}}} \right| \tag{3}$$

The null hypothesis of no change is rejected if $D_n > h_n$, where hn is the critical value (threshold) chosen for the test. The critical values of D_n for a range of test statistics have been generated using simulation methods by [20]. These computed values are employed in the R package cpm developed by [19]. The best estimate of the location of the change point will be the observation that immediately followed the value of k which maximised D_n.

In this study there are over 100,000 observations, so nonparametric tests for change in location and sale will be used. Two nonparametric statistics will be employed, the Mood statistic for change in volatility (scale) and a Lepage type statistic which tests for a change in location and scale, the results of which are presented in Fig. 2. The implementation of these statistics for change point detection in the R cpm package were used to establish the existence and location of a change point in the bitcoin price series.

Both the Mood and Lepage statistics indicate there is a significant change in the distribution, driven by the increase in volatility. The date of the change is 29 November 2017, two days before the official announcement of the commencement dates for futures trading. As returns for financial assets have often been found to be non i.i.d. the analysis was run on the raw returns and residuals from a

ARMA(1,1)-GARCH (1,1)[2] model fit. A significant change in the distribution, associated with the increase in series volatility was detected at the same point in time in each case. The increase in volatility will exacerbate the issue of bitcoin not being a reliable store of value.

The next part of our analysis will measure the extent of risk reduction that can be obtained by forming hedge portfolios. One of the uses of futures contracts is to manage (hedge) pricing risk in the underlying spot market. That is, by taking offsetting positions in the spot and futures markets risk, as measured by return volatility, can be reduced. In practice this means that the volatility of the hedge portfolio, a combination of the underlying asset and futures contracts, will be less than the volatility of the unhedged position in the spot market. It is possible that an appropriately constructed hedge portfolio can be used to manage the volatility of bitcoin prices. The body of research which has looked at the calculation and evaluation of optimal hedging strategies, such as [5,6,11,16], concludes that hedge ratios selected by OLS generally work best when evaluated in sample. We will analyse naive and OLS based hedging strategies.

The effectiveness of the hedge can be measured by the percentage reduction volatility that results from holding the hedge portfolio:

$$\text{Effectiveness} = 1 - \frac{\text{Variance}_{hp}}{\text{Variance}_{up}} \tag{4}$$

$$\text{Risk Reduction} = \frac{\sigma_{up} - \sigma_{hp}}{\sigma_{up}} \tag{5}$$

where hp and up represent a hedged portfolio and unhedged portfolio respectively. These metrics will be positive only if there has been a reduction in risk, the volatility of the hedged position is less than the volatility of the unhedged position. We will also compute Hedge Effectiveness using Semi-variance, which measures the variability of returns below the mean, addressing a shortcoming of the variance and providing a more intuitive measure of risk for hedging focusing on downside risk.

Two hedging approaches are evaluated. The first is the naive hedge which is a portfolio with one short futures position for every bitcoin position. The return of this naive hedge portfolio is the spot return minus the futures return. The second approach is the ordinary least squares hedge (a form of optimal hedge). A simple OLS regression of the form $r_{spot} = \alpha + \beta_{future}$ is run. The estimated β is used as the hedge ratio when calculating the hedge portfolio return,

---

[2] The ARMA(1,1) model has the following form: $\Delta P_t = \alpha_0 + \beta_1 \Delta P_{t-1} + \beta_2 \Psi_{t-1} + \Psi_t$, while the GARCH(1,1) model specification also considers $\sigma^2 = \alpha_1 + \gamma_1 \Psi_{t-1}^2 + \gamma_2 \sigma_{t-1}^2$ where the conditional variance term (σ^2) is the one-period ahead forecast variance based on past information and is a function of three terms: the mean; news about volatility from the previous period, measured as the lag of the squared residual from the mean equation (the ARCH term $\gamma_1 \Psi_{t-1}^2$); and last period's forecast variance (the GARCH term $\gamma_2 \sigma_{t-1}^2$). This specification interprets this period's variance as being formed by a weighted average of a long-term average (the constant), the forecast variance from the last period (the GARCH term), and information about volatility observed in the previous period (the ARCH term).

$r_{HedgedPortfolio} = r_{spot} - \beta r_{future}$. This approach to hedging is implemented using a rolling regression framework. The β is estimated on a sample of observations and then is used out of sample in the computation of the hedge portfolio return. In this work the β is estimated each day then used to compute the hedge portfolio return for the next day. The return series for the dynamic hedge is the concatenation of each days computed hedge portfolio return. Table 3 contains the results of the evaluation of hedge effectiveness.

Table 3. Hedge effectiveness

Naive Hedge	
Risk reduction	−334.59
Hedge effectiveness	−3.3459
Hedge effectiveness (semivarance)	−1.20851
Rolling OLS Hedge	
Risk reduction	−60.7627
Hedge effectiveness	−0.60763
Hedge effectiveness (semivarance)	−0.38919

The first and most striking result is that hedging increases risk, as indicated by the negative effectiveness and risk reduction results. The variance of the hedged portfolio is greater that the variance of the unhedged position under both hedging strategies. While the rolling OLS hedge is more effective than the naive hedge, as would be expected, it also increases the pricing risk inherent in holding physical bitcoin. Using semi-variance in the computation of hedge effectiveness, thus only focusing on downside risk, shows an improvement in effectiveness compared to the use of the variance. However both the hedging strategies are shown to be risk increasing under all evaluation methods.

It is generally accepted that futures contracts lead their respective underlying assets in price discovery, [1,3,10,18]. These results highlight the importance of market structure and instrument type. The findings of these studies indicate that the centralisation and relative transparency of futures markets contribute to their large role in price discovery. It is also likely that low transaction costs, inbuilt leverage, ease of shorting and the ability to avoid holding the underlying physical asset make futures contracts an attractive alternative for traders in a wide range of assets. [1] argue that the emergence of futures markets generally coincides with the rise of instructional trading. The trades of sophisticated institutional investors contributes to price discovery being focused in futures markets.

There are two measures of price discovery commonly employed in the literature, the [9] Information Share (IS) and the [7] Component Share (CS). [9] demonstrates that the contribution of a price series to price discovery (the 'information share') can be measured by the proportion of the variance in the common efficient price innovations that is explained by innovations in that price series.

[7] decompose a cointegrated price series into a permanent component and a temporary component using error correction coefficients. The permanent component is interpreted as the common efficient price, the temporary component reflects deviations from the efficient price caused by trading fractions. We estimate IS and CS, as developed by [10] using the error correction parameters and variance-covariance of the error terms from the Vector Error Correction Model (VECM):

$$\Delta_{p1,t} = \alpha_1(p_{1,t-1} - p_{2,t-1}) + \sum_{i=1}^{200} \gamma_i \Delta p_{1,t-i} + \sum_{j=1}^{200} \delta_j \Delta p_{2,t-j} + \varepsilon_{1,t} \quad (6)$$

$$\Delta_{p2,t} = \alpha_2(p_{1,t-1} - p_{2,t-1}) + \sum_{k=1}^{200} \varphi_k \Delta p_{1,t-k} + \sum_{m=1}^{200} \phi_m \Delta p_{2,t-m} + \varepsilon_{2,t} \quad (7)$$

where $\Delta p_{i,t}$ is the change in the log price $(p_{i,t})$ of the asset traded in market i at time t. The next stage is to obtain the component shares from the normalised orthogonal to the vector of error correction coefficients, therefore:

$$CS_1 = \gamma_1 = \frac{\alpha_2}{\alpha_2 - \alpha_1}; CS_2 = \gamma_2 = \frac{\alpha_1}{\alpha_1 - \alpha_2} \quad (8)$$

Given the covariance matrix of the reduced form VECM error terms [3] where:

$$M = \begin{pmatrix} m_{11} & 0 \\ m_{12} & m_{22} \end{pmatrix} = \begin{pmatrix} \sigma_1 & 0 \\ \rho\sigma_2 & \sigma_2(1-\rho^2)^{\frac{1}{2}} \end{pmatrix} \quad (9)$$

we calculate the IS using:

$$IS_1 = \frac{(\gamma_1 m_{11} + \gamma_2 m_{12})^2}{(\gamma_1 m_{11} + \gamma_2 m_{12})^2 + (\gamma_2 m_{22})^2} \quad (10)$$

$$IS_2 = \frac{(\gamma_2 m_{22})^2}{(\gamma_1 m_{11} + \gamma_2 m_{12})^2 + (\gamma_2 m_{22})^2} \quad (11)$$

Recent studies show that IS and CS are sensitive to the relative level of noise in each market, they measure a combination of leadership in impounding new information and the relative level of noise in the price series from each market. The measures tend to overstate the price discovery contribution of the less noisy market. An appropriate combination of IS and CS cancels out dependence on noise, [17,21]. The combined measure is known as the Information Leadership Share (ILS) which is calculated as:

$$ILS_1 = \frac{\left| \frac{IS_1}{IS_2} \frac{CS_2}{CS_1} \right|}{\left| \frac{IS_1}{IS_2} \frac{CS_2}{CS_1} \right| + \left| \frac{IS_2}{IS_1} \frac{CS_1}{CS_2} \right|} \text{ and } ILS_2 = \frac{\left| \frac{IS_2}{IS_1} \frac{CS_1}{CS_2} \right|}{\left| \frac{IS_1}{IS_2} \frac{CS_2}{CS_1} \right| + \left| \frac{IS_2}{IS_1} \frac{CS_1}{CS_2} \right|} \quad (12)$$

We estimate all three price discovery metrics, noting that they measure different aspects of price discovery.

---

[3] $\Omega = \begin{pmatrix} \sigma_1^2 & \rho\sigma_1\sigma_2 \\ \rho\sigma_1\sigma_2 & \sigma_2^2 \end{pmatrix}$ and its Cholesky factorisation, $\Omega = MM'$.

Table 4. Price discovery results

Information Share (Hasbruck)	Lower Bound	Upper Bound	Average
Futures	0.115535	0.183738	0.149637
Bitcoin	0.816261	0.884465	0.850363
Component Share (Gonzalo)	Average		
Futures	0.177028		
Bitcoin	0.822971		
Information Leadership (Yan)	Average		
Futures	0.025636		
Bitcoin	0.827931		
Information Leadership Share (Putnins)	Average		
Futures	0.030034		
Bitcoin	0.969965		

The results in Table 4 show that the spot market leads in price discovery according to all the metrics computed. This result is contrary to what has been found in a range of other asset classes, where futures markets lead. Looking at the Information Leadership Share; 97% of the information affecting bitcoin prices is reflected in the spot market, the remaining 3% is reflected in the futures market. The concentration of price discovery in the spot market may be a function of the novelty of the new futures contracts, they have been trading for 3 months. It may also be the case that the type of investor attracted to bitcoin because of its anonymity may not be inclined to begin trading on a registered and regulated futures market where personal details have to be given before trading is permitted. These investors would in general be classified as uninformed. Because of various restrictions on bitcoin there is an absence of a large cohort of institutional investors who have positions in physical bitcoin. The results presented support the argument put forward by Bohl et al. that the dominance of unsophisticated individual investors in the futures market impedes its contribution to price discovery.

4   Conclusions

The economic attributes of a currency are; it is a medium of exchange, a store of value and a unit of account. [22] asserted that bitcoin was not a currency as it 'performs poorly as a unit of account and as a store of value'. The high volatility of bitcoin prices and the range of prices quoted on various bitcoin exchanges were seen to damage bitcoin's usefulness as a unit of account. If the introduction of trading in bitcoin futures resulted in a reduction in the variance of bitcoin prices, or facilitated hedging strategies that could mitigate pricing risk in the spot market it is possible that bitcoin could act as a unit of account, moving it closer to being a currency. The analysis conducted shows that volatility increased around the announcement of trading in bitcoin futures. In the period covered by this study hedge portfolios constructed with the futures cannot mitigate the risk inherent in the underlying spot market, both of the hedging strategies considered resulted in an increase in volatility. The price discovery analysis indicated that price discovery is focused on the spot market, which is in keeping with the argument that the traders in the futures market are uninformed noise traders. Together these results support Yermak's conclusion that bitcoin should be seen as a speculative asset rather than a currency.

References

1. Bohl, M.T., Salm, C.A., Schuppli, M.: Price discovery and investor structure in stock index futures. J. Futures Mark. **31**(3), 282–306 (2011)
2. Böhme, R., Christin, N., Edelman, B., Moore, T.: Bitcoin: economics, technology, and governance. J. Econ. Persp. **29**(2), 213–38 (2015)
3. Cabrera, J., Wang, T., Yang, J.: Do futures lead price discovery in electronic foreign exchange markets? J. Futures Mark. **29**(2), 137–156 (2009)
4. Cheah, E.-T., Fry, J.: Speculative bubbles in bitcoin markets? an empirical investigation into the fundamental value of bitcoin. Econ. Lett. **130**, 32–36 (2015)
5. Choudhry, T.: Short-run deviations and optimal hedge ratio: evidence from stock futures. J. Multinational Financ. Manage. **13**(2), 171–192 (2003)
6. Figlewski, S.: Hedging performance and basis risk in stock index futures. J. Financ. **39**(3), 657–669 (1984)
7. Gonzalo, J., Granger, C.: Estimation of common long-memory components in cointegrated systems. J. Bus. Econ. Stat. **13**(1), 27–35 (1995)
8. Gulen, H., Mayhew, S.: Stock index futures trading and volatility in international equity markets. J. Futures Markets Futures Options Other Deriv. Prod. **20**(7), 661–685 (2000)
9. Hasbrouck, J.: One security, many markets: determining the contributions to price discovery. J. Financ. **50**(4), 1175–1199 (1995)
10. Hauptfleisch, M., Putniņš, T.J., Lucey, B.: Who sets the price of gold? London or New York. J. Futures Markets **36**(6), 564–586 (2016)
11. Kroner, K.F., Sultan, J.: Time-varying distributions and dynamic hedging with foreign currency futures. J. Financ. Quant. Anal. **28**(4), 535–551 (1993)
12. Lee, C.L., Stevenson, S., Lee, M.-L.: Futures trading, spot price volatility and market efficiency: evidence from european real estate securities futures. J. Real Estate Financ. Econ. **48**(2), 299–322 (2014)
13. Lepage, Y.: A combination of wilcoxon's and ansari-bradley's statistics. Biometrika **58**(1), 213–217 (1971)
14. Mood, A.M.: On the asymptotic efficiency of certain nonparametric two-sampletests. Ann. Math. Stat. **25**, 514–522 (1954)
15. Nakamoto, S.: Bitcoin: A Peer-to-Peer Electronic Cash System (2008)
16. Park, T.H., Switzer, L.N.: Bivariate garch estimation of the optimal hedge ratios for stock index futures: a note. J. Futures Markets **15**(1), 61–67 (1995)
17. Putniņš, T.J.: What do price discovery metrics really measure? J. Empirical Financ. **23**, 68–83 (2013)
18. Rosenberg, J.V., Traub, L.G.: Price discovery in the foreign currency futures and spot market. J. Deriv. **17**(2), 7–25 (2009)

19. Ross, G.J., et al.: Parametric and nonparametric sequential change detection in r: the cpm package. J. Stat. Softw. **66**(3), 1–20 (2015)
20. Ross, G.J., Tasoulis, D.K., Adams, N.M.: Nonparametric monitoring of data streams for changes in location and scale. Technometrics **53**(4), 379–389 (2011)
21. Yan, B., Zivot, E.: A structural analysis of price discovery measures. J. Financ. Markets **13**(1), 1–19 (2010)
22. Yermack, D.: Is bitcoin a real currency? an economic appraisal. In: Handbook of Digital Currency, pp. 31–43. Elsevier (2015)

Blockchained *Sukuk*-Financing

Shazib Shaikh[✉] and Fatima Zaka

Information Technology University,
Arfa Software Technology Park, Lahore, Pakistan
{shazib.ehsan, fatima.zaka}@itu.edu.pk

Abstract. *Sukuk* have proven to be a significant innovation in the Islamic finance industry, introduced as an alternative to conventional bonds and securities. Aiming for fairer distribution of market risk between the haves and have-nots, Islamic finance focuses on funding the purchase of real assets rather than simply borrowing money. However, in reality, this requirement of "asset-backed" borrowing has limited the growth of the *sukuk* segment. We propose the application of blockchain technology to enhance the traceability of credit to the specific financed assets. *Sukuk* transactions can involve a number of parties, especially when sale, lease and agency contracts are combined. Some recent *sukuk*-default scandals transpired to have avoided "true sale" of underlying assets. Since the opinion of religious scholars can have a significant impact on the valuation of *sukuk*, we believe that the traceability of asset transfers will enhance *sukuk* credibility and valuation. Moreover, a smart contract infrastructure with blockchain security should also significantly reduce the execution time for such transactions. We present an initial model for blockchained *sukuk*-issue in this paper, highlighting the significant design features that specifically concern this niche market.

Keywords: Blockchain · Fintech · Islamic finance · Sukuk · Smart contract

1 Introduction

The societal disharmony that comes with a large gap between rich and poor is essentially what Islamic models of financing aim to alleviate. By eyeing a fairer distribution of risk and return between a lender and borrower, these models emphasize buying assets on credit rather than taking loans in cash.

We believe that blockchain technologies can greatly facilitate the type of large-scale asset-funding required by modern businesses and developers of infrastructure. By enabling better traceability of funds against specific assets, it should appeal to a significant segment of the lending market – one that is religiously conscientious. In this conceptual paper, we specifically look at blockchaining the Islamic alternative to bonds and securities called *sukuk* (pronounced *su-kook'* – plural of *sak*).

The remainder of this paper is structured as follows. Section 2 provides a necessary background on the significance and nature of Islamic finance and, within it, *sukuk*. Section 3 introduces relevant aspects of blockchain technology. Section 4 evaluates the utility of blockchain for *sukuk* investors. Section 5 presents our conceptual model for block-chained *sukuk*. Lastly, we discuss future avenues of research in Sect. 6 and provide overall conclusions in Sect. 7.

© Springer Nature Switzerland AG 2019
N. Mehandjiev and B. Saadouni (Eds.): FinanceCom 2018, LNBIP 345, pp. 66–76, 2019.
https://doi.org/10.1007/978-3-030-19037-8_5

2   Background and Trends in Islamic Finance

Islamic finance is a significant niche in today's financial markets. The following key statistics and trends should highlight this (as reported in the Islamic Financial Services Industry Stability Report in 2017 [9]):

1. The total worth of the Islamic Financial Services Industry (IFSI) in 2016 was estimated to be USD 1.89 trillion (in 2014, it was at USD 1.87 trillion – at a time when assets of the top 1000 global conventional banks declined 2.6%).
2. Out of this, USD 318.5 billion were the outstanding *sukuk* that registered a 6.06% expansion (the largest component of IFSI is Islamic banking with USD 1.493 trillion in global banking assets which showed a slight decline).
3. In the 2nd quarter of 2016, as a percentage of total banking assets, market share of Islamic banking assets in Iran, Brunei, Saudi Arabia, Malaysia and the UAE were 100%, 100%, 51%, 28% and 25% respectively.
4. Around 80% of *sukuk* issuances in 2016 were by national governments (sovereign *sukuk*). Of these, more than 50% were by Malaysia, followed by Indonesia, UAE and Saudi Arabia. In UAE, the issuers were notable entities such as Etihad Airways and Dubai DP World Crescent Ltd., while in Saudi Arabia, the Islamic Development Bank makes a regular annual issue.

In keeping with the tradition of other Abrahamic faiths, Islam stipulates a strong prohibition on interest-based earnings. The legal maxim, attributed to the final Prophet of Islam, Muhammad (peace be upon him), states that "profit comes with liability" - thus requiring a lender to share the risk of the investment [11]. However, Islamic jurists do allow for the sale of real goods on credit at a higher price (vis-à-vis cash sales). In essence, this is the foundation for most of the modern Islamic financing models, including *sukuk* [11, 18]. However, when positioned against conventional borrowing, such sales are often described as "asset-backed" borrowing. The other basic model of Islamic finance institutes partnerships where the lender is a sleeping partner and there exists profit-and-loss sharing.

Whether credit-sale or partnership, under such a regime, the lender now faces more exposure to loss as their risk now extends beyond the borrower to the underlying investment itself. Thus, IFSI have often combined their products with parallel rental agreements (called *Ijarah* or operating leases). The receiver of funds retains immediate usage of the asset, paying rent on the share that they do not own, until they have bought back the whole asset at maturity. Such sale-with-lease issuances are a significant component of the *sukuk* market. International Islamic Financial Market (IIFM) estimated that these issuances constituted around 35% of the international *sukuk* market share in 2010–2015. But this declined to 16% in 2016 [10].

This displacement is explained by the growth of a specific form of partnership-based *sukuk* called *wakalah sukuk*. According to IIFM, these type of *sukuk* increased from 43% of international issuances in 2010–2015 to 75% in 2016 [10]. IIFM does concede that the *Wakalah* contracts that they reported include an element of *Ijarah* within them. The agreement is no longer focused on assets; rather, the funding-provider now employs an agent (which is usually either the fund-raiser or a related party) to oversee a pool of assets. As an incentive for lucrative asset management, from the profits, what is over and above a certain profit level is kept by the agent.

Ultimately, the target market for IFSI products are conscientious investors who have surplus funds and who wish to avoid conventional interest-bearing investments. Otherwise, in certain Muslim majority economies, these funds remain in the informal sector in abstinence from the banking sector [1].

3   Blockchain: A Secure and Distributed Ledger

"A blockchain is essentially a distributed database of records, or public ledger of all transactions or digital events that have been executed and shared among participating parties. Each transaction in the public ledger is verified by consensus of a majority of the participants in the system." [4, p. 7]

While it is the popularity of cryptocurrencies such as Bitcoin that have brought blockchain technology to the fore, the definition above clarifies that the blockchain is a digital ledger that can keep record of *any* information that has any value. When a transaction takes place, it is added to a chain of blocks on multiple machines in this network of ledgers. As described by some: "Each block becomes an independent banking statement whereas a blockchain is a complete history of all banking transactions." [19, p. 7]

Once a pre-specified set of transactions is completed, the blocks are encapsulated in a hashed blockchain. Each block that extends the chain results in a new hash value, effectively placing a lock atop of a previous lock. In this manner, the transactions in a block become "immutable", meaning that they are not only irreversible, but that they can also not be tampered with due to the prohibitive computing power required.

The blockchain could be public, private or could even be based on an open source protocol. A public blockchain platform, such as that of Bitcoin and Ethereum, makes accessing a transaction in a block as easy as accessing any other resource on the Internet [23]. Ethereum actually goes beyond simply tracking cryptocurrency transactions and enables implementation of "smart" contracts in a distributed computing framework [22]. Therefore, contracts can be automatically executed and enforced without the involvement of third-parties that would otherwise be needed in their design and execution.

Credibility of each party in a transaction may optionally be further protected through well-established public key cryptography. Due to distributed validation and strongly secured record-keeping, the need for active involvement of intermediaries to closed agreements and broker trust (such as commercial banks, central banks and other agencies) is reduced. It is reported that consumers could save up to USD 16 billion in banking and insurance fees each year through the use of blockchain-based applications [17].

4   The Value of Blockchain to *Sukuk* Investors

4.1   Value Shared with Conventional Bonds – but Multiplied

Based on the relative ease-of-access, distributed validation and robust security mentioned in the last section, some have already recommended blockchain for even conventional bonds. The typical problems this should remedy are as follows [19, p. 5–6]:

1. Time-consuming resolution of data inconsistencies usual among multiple parties.
2. Elongated clearing and settlement times, especially due to intermediaries.
3. Risks due to delays that occur between advance payments and issuance of security.
4. Multi-step processes that are prone to greater errors.

The value expected from blockchaining *sukuk* can be multiplied, given we can add the following specific complications to this list:

5. Traceability to the underlying asset is a key criteria for legitimizing *sukuk* over bonds with Islamic jurists.
6. Greater time taken from decision-to-issue till go-to-market for *sukuk*. For conventional bonds, it can take days, but *for sukuk* it can take from "a few weeks to a few years" (a key reason given for the decline in corporate *sukuk*) [9, p. 22].

We elaborate on these peculiar aspects of *sukuk* in the next two sub-sections.

4.2   Traceability: A Religious Requirement

Some *sukuk* defaults in the earlier part of this decade highlighted the need to recognize the prevalence of the "asset-*based*" *sukuk* model – a compromised form of the ideal "asset-backed" *sukuk* model [16]. It came to the fore that, instead of selling complete ownership of the underlying asset, the originators had only sold the "beneficial ownership" in the asset-based model. The *sukuk*-holders could gain profit from the asset. However, at the time of default, they could not execute their resale (much like a conventional trust). In fact, during the 2008 global financial crisis, the *sukuk* that defaulted were all asset-based and not asset-backed [16]. In spite of this, asset-backed *sukuk* still remain a small component of *sukuk* issuances due to the strong preference of the originator to retain practical ownership of the pledged assets [16].

The legitimacy of *sukuk* among reputable Islamic jurists has a significant impact on the valuations of *sukuk* [8]. Furthermore, beyond the requirement of interest-avoidance and risk-sharing, Islamic financing is also explicitly meant to reduce uncertainty and speculative behavior [11]. Hence, gambling and short-selling (selling an asset you do not yet own) are prohibited under Islamic law.

As a result, we believe that by enhancing traceability of funds to specific registered assets, credibility and tradability of *sukuk* will improve among investors who have avoided them thus far. It addresses a skepticism that persists that Islamic financing is "a lion in sheep's skin"; that *sukuk* are really conventional bonds with Islamic models being "grafted" upon them to fit models of finance in vogue [12].

A greater need for traceability of *sukuk* assets can arise when there is actually a *pool* of assets, instead of a set of easily identifiable and countable ones. Moreover, there has been an appetite to grant flexibility to the asset-seller (who then leases them back) in substituting assets within the pool prior to *sukuk* maturity. This is desirable when the issuance concerns a property developer who would like to sell some real estate from within the pool. This flexibility can be granted to the issuer using block chained traceability [2]. Additionally, for the buyer of the *sukuk*, selling their asset holdings can be more transparently carried out. This encourages greater liquidity in *sukuk* markets.

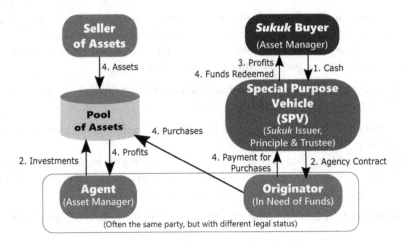

Fig. 1. Complexity of roles in an agency-based (*wakala*) *sukuk* structure

4.3 Complexity Needing Transparency and Disintermediation

Sukuk issuance can, in reality, take a highly complex legal form, as it often involves drawing up parallel contracts of sale, rental and agency between multiple parties. This is especially the case in *wakalah sukuk* (agency-based *sukuk*), where up to three or four parties can come between the *sukuk* buyer and the underlying pool of assets being managed on their behalf (see Fig. 1). Then there are the ancillary parties (e.g. lawyers, corporate bankers, etc.) who normally facilitate the set-up of such a venture. Therefore, we believe that traceability, brought especially by this kind of a block-chained distributed ledger, will greatly enhance investor confidence by overcoming complexity.

Implementation on a public platform such as Ethereum supports self-executing contracts, hashed transactions and secure digital sign-offs by the parties and regulating institutions. This should reduce the need for intermediaries, especially when governments standardize templates of such self-enforcing contracts.

The next section illustrates the form that *sukuk* may take when implemented on a blockchain platform.

5   Conceptual Model for Blockchained *Sukuk*

5.1   Blockchain Design Issues for *Sukuk* Management

In adopting blockchain for the securities market, Wall and Malm [20] identified a number of key design issues, which we enumerate as follows:

1. What contracts will be converted to smart contracts on the blockchain platform?
2. Will it be a permissioned or permissionless distributed ledger?
3. How will "delivery vs payment" (DvP) be implemented? That is, how will asset ownership be tracked, and how will the transfers be executed and verified?

Our conceptual model for blockchained *sukuk* will focus on these design issues as they are sufficient to deliver the benefits we have envisioned.

5.2 Tracking Contracts and Asset Transfers with a Triplicate Ledger Model

The set of contracts constituting a *Sukuk* issuance affect three types of transfers that will need to be tracked separately. These are as follows:

1. the transfer of each asset underlying the *sukuk* agreement,
2. the transfer of *sukuk* that each lay claim to only a share of the asset(s) and
3. the transfer of cash or other currency in which the asset(s) has been tokenized.

Fig. 2. *Sukuk* model in which assets are sold and then leased back (*ijarah sukuk*)

To illustrate how *sukuk* issuance can be blockchained, let us take the example of a common *ijarah sukuk* that is structured as a credit sale followed by a lease-back agreement (see Fig. 2). The party requiring the funds that originally owns the assets is called the Originator. To manage the *sukuk* as a trust, the creation of a Special Purpose Vehicle (SPV) is a standard procedure. It is the SPV that buys (or even leases) the assets being pledged against the *sukuk,* and then rents it back to the Originator so that the SPV can generate a regular income for the *sukuk* buyers.

When implemented in a blockchain, we assume a three ledger model to track the three items of interest [7, 20, Ch. 3] (see Table 1). Note that we consider the term "ledger" synonymous with a wallet or registry in Bitcoin or Ethereum terminology. We do not constrain the precise choice of blockchain platform at this stage. All three registries may be implemented within the same blockchain, or the cash/token ledger may be assumed to exist on a separate blockchain, such as Bitcoin or Ethereum [20].

For the rest of this paper, we will assume that each of the smart contracts is implemented within Ethereum, as it already enables storing the state of smart contracts, (i.e. the extent to which the code has been executed and the state of the objects it creates as a result) in addition to storing the transaction itself (such as on Bitcoin). It also provides its own cryptocurrency (Ether) that can be used to tokenize assets.

However, the choice of Ethereum implies certain features which may need to be varied depending on the jurisdiction in which the *Sukuk* contracts originate. Smart contracts are hosted on each node in a distributed computing architecture. Each node receives Ethers for maintaining the state of the contract regardless of the application that is implemented in this distributed manner. To the node, the smart contract is a

Table 1. A blockchained *ijarah sukuk* transaction modeled with triplicate ledgers

Smart contract	Asset ledger	Sukuk ledger	Token ledger
1. Asset tokenization	Originator registers asset and creates token value for asset (National 3rd Party may digitally certify asset)	N/a	N/a
2. Sale of assets to SPV	Transfer ownership to SPV from Originator (in escrow until cash delivery)	N/a	Transfer of Cash to Originator from SPV at agreed date (in escrow until asset delivery)
3. Issuance of *Sukuk*	SPV transfers asset ownership, creating public key for each owner who has a *Sak*	Creation and issuance of *Sukuk* (National 3rd Party may digitally certify)	Transfer of Cash/Token from Sukuk buyers to SPV
4. Rent back of asset	Some parameters (except ownership), such as asset location and condition may be updated	Dividends registered against *Sukuk*	Transfer of Rents from Originator to individual *Sukuk* holders
5. Buy back of asset	Transfer of ownership to Originator from SPV	*Sukuks* redeemed and ownership claims terminated	Transfer of buy back cash/tokens to *sukuk* holders

black box, and it is only concerned with mining the transaction state changes and hashing them as part of a block. Consensus regarding the valid state of the block (like the Bitcoin blockchain) is by providing "Proof-of-Work" [23]. This means that the node competes to solve a computationally complex mathematical puzzle, linked to the validity of the transaction state, in order to earn Ethers.

Trust is created as this puzzle-solving occurs in parallel on several nodes, and the computational energy required to tamper with any block is prohibitive. However, as this "Proof-of-Work" method of consensus is itself quite energy-intensive and limits how many transactions are processed as part of a block, the Ethereum platform is planning to move to "Proof-of-Stake" consensus [23]. That would only require miners to provide ownership of currency (their stake in the blockchain platform) as evidence of their vested interest in the blockchain. This is apparently more scalable.

Note that by specifying the smart contract transactions in Table 1, we are not restricting how many transactions will be bundled into a block. That is a feature of the blockchain platform itself. The immutability of an asset transfer is not harmed by bundling multiple transactions into a single block [23].

5.3 Permissioned Ledgers?

A platform such as Bitcoin is unconcerned with the identities of its users as it encapsulates this in alphanumeric addresses contained in public and private keys. However, current financial regulatory climate requires explicit identification of participating entities [20, p. 59–60]. Hence, the synopsis of a European Central Bank paper on distributed ledger technologies stated that "certain processes that feature in the post-trade market for securities will still need to be performed by institutions" [15, p. 2].

Ethereum does provide for consortia to create their own permissioned ledgers within its blockchain eco-system [23]. The focus of our blockchained *sukuk* initiative is to assure legal and complete transfer of assets to the correct parties. We can assume trust is not lacking in the national governments conducting asset transfers. Rather it is in the originator and SPV: Did they fully transfer ownership to the *sukuk*-holders or not? Therefore, a permissioned ledger solution with digital certification by trusted institutions should be viable in principle.

Nevertheless, certain governments, such as that of Georgia [21] and India [14], are experimenting with direct use of the Bitcoin blockchain as a parallel system to the existing offline registration process, especially in the area of land registration. Hence, such a parallel ledger model, which still involves the government as the creator of this blockchain layer, appears viable and acceptable. It also has the advantage of exploiting mining facilities of existing blockchain platforms.

But for *sukuk*, involvement of regulatory bodies is quite essential, if legitimacy is indeed the objective. There needs to be registration of the asset ownership with a legal authority, perhaps various government agencies, and the registration of the *sukuk* listing with a securities regulator (such as the Securities and Exchange Commission). Conventional bonds may register collateral, but it is not necessary to comply with this registration where "bearer bonds" are concerned [19]. Hence, permissioned ledgers within the Ethereum eco-system appears to most fit *sukuk* blockchain requirements.

5.4 Delivery vs. Payment

This is an area where blockchain actually facilitates asset transfers more efficiently and effectively than traditional methods. Essentially, if the whole transaction is carried out on a platform such as that of Bitcoin, "partial transactions" are created and the cash tokens are placed in escrow until the delivery versus payment is received [20, Ch. 3]. The transaction remains contingent upon a combination of public/private-key sign-offs from involved parties.

As Ethereum is more geared towards enabling smart contracts that can maintain and operate on multiple object registries or wallets simultaneously, the asset recording mechanism is much more developed. Already functioning asset tokenization implementations on Ethereum such as Digix (allowing conversion of assets into gold bullion tokens) [6], and LAToken Ethereum Smart Contracts (allowing fractions of real estate to be traded) [13], as well as standards such as ERC-721 Non-Fungible Token Standard [5] have been finalized. They go beyond the management of the intermediate escrow state and discuss transfers of fractions of assets.

Hence, beyond electing a platform such as Ethereum, there is little more value we can add to this discussion. Tokenization of the assets and the *sukuk* is supported by the state-of-the-art as we envision in Table 1. Therefore, it is for the concerned parties to decide which option suits their business model best. For example, the Digix "Gold Standard" may actually appeal to some Islamic Finance audiences, as some academics argue for monetary systems that are based on fully gold-backed currencies [3].

6 Way Forward

6.1 Other *Sukuk* Models?

The previous section has only demonstrated how the sale-with-lease *sukuk* (*ijarah sukuk*) can be converted to a three-ledger model. This is easily generalizable to other models of *sukuk* as it covers the core transactions involving asset-ownership transfers in our proposed "Asset Ledger" stream (refer back to Table 1). The other forms of *sukuk* financing are either reducible to the simple "asset-sale with buy-back", or add an extra contract of agency, as in the case of the popular *wakalah sukuk* (introduced earlier in Fig. 1). The latter involves no asset transfer. However, a question arises when the agency involves development of a property. In this case, accounting conventions and standards will have to be agreed on between the parties, after which recording in the blockchained asset ledger can be executed accordingly.

6.2 Implementation and Evaluation

Another important avenue for further work is the implementation of a proof-of-concept *sukuk* blockchain. A decision needs to be made between an Ethereum hosted block-chain or a custom permissioned ledger that involves evaluation by one or more national regulators. A further aspect requiring finalization is the extent to which existing asset tokenization software (e.g. LAToken or Digix) are practically conducive to *sukuk* blockchaining requirements.

The first stage of evaluation will look at technical feasibility by a proof-of-concept prototype: can the model conceptualized actually be operationalized as faithfully as expected and in a manner acceptable to a panel of regulatory authority and security experts. The second will have to evaluate market feasibility, probably through a survey of feedback from Islamic scholars of the prototype, and simulated *sukuk* trading with representative panel of stakeholders.

7 Conclusion

We have highlighted the need for the credibility of *sukuk* in the context of a historic practice of offering them as "asset-based" rather than "asset-backed". As we have seen, *sukuk* transactions involve a number of entity layers coming between the *sukuk* buyer and their asset – which can often allow incomplete asset sales and restricted "beneficial ownership" sales only. With blockchain, we have not only proposed an efficient and cost-saving solution, but also one that goes to the core of the legitimacy of the product. Given that legitimacy among Islamic scholars improves the viability of Islamic Finance products, we have argued that blockchained *sukuk* should in principle attract more mainstream Islamic Finance investors. To that end, we have presented a three-ledger conceptual model for blockchain implementation that should ensure a minimum viable asset transfer record. We foresee the future steps as experimentation with an open blockchain protocol such as Ethereum, to allow for further optimization of the blockchain set-up, and to evaluate acceptability amongst regulating bodies.

References

1. Abedifar, P., Hasan, I., Tarazi, A.: Finance-growth nexus and dual-banking systems: relative importance of Islamic banks. J. Econ. Behav. Organ. **132**, 198–215 (2016)
2. Al-Amine, M., Al-Bashir, M.: Sukuk market: innovations and challenges. Islamic Econ. Stud. **15**, 1–22 (2008)
3. Choudhury, M.A.: Islamic dinar and 100 percent reserve requirement monetary system. Int. J. Manag. Stud. **15**, 1 (2008). http://ijms.uum.edu.my/images/pdf1/15no2ijms/ijms1521.pdf. Accessed 24 July 2018
4. Crosby, M., Pattanayak, P., Verma, S., et al.: Blockchain technology: beyond bitcoin. Appl. Innov. **2**, 6–10 (2016)
5. Entriken, W., Shirley, D., Evans, J., et al.: ERC-721 Non-Fungible Token Standard. Ethereum Improvement Proposals (2018). https://github.com/ethereum/EIPs/blob/master/EIPS/eip-721.md. Accessed 24 July 2018
6. Eufemio, A.C., Chng, K.C., Djie, S.: Digix's Whitepaper: The Gold Standard in Crypto Assets. White paper (2016). https://www.weusecoins.com/assets/pdf/library/Digix%20Whitepaper%20-%20The%20Gold%20Standard%20in%20Crypto%20Assets.pdf. Accessed 24 July 2018
7. Eze, P., Eziokwu, T., Okpara, C.: A Triplicate Smart Contract Model using Blockchain Technology. Circulation in Computer Science, Special Issue on Disruptive Computing, Cyber-Physical Systems (CPS), and Internet of Everything (IoE), pp. 1–10 (2017). https://doi.org/10.22632/ccs-2017-cps-01
8. Godlewski, C.J., Turk-Ariss, R., Weill, L.: Do the type of sukuk and choice of shari'a scholar matter? J. Econ. Behav. Organ. **132**, 63–76 (2016)
9. IFSB: Islamic Financial Services Industry Stability Report 2017. Islamic Financial Services Board (2017)
10. IIFM: IIFM Annual Sukuk Report. International Islamic Financial Market (2018)
11. Iqbal, Z., Mirakhor, A.: An Introduction to Islamic Finance: Theory and Practice. Wiley, Hoboken (2011)
12. Lai, J., Rethel, L., Steiner, K.: Conceptualizing dynamic challenges to global financial diffusion: Islamic finance and the grafting of sukuk. Rev. Int. Polit. Econ. **24**, 958–979 (2017)
13. LAToken: Tokens Smart contracts of the LAToken Project (https://latoken.com). GitHub (2018). https://github.com/LAToken/eth-smart-contracts. Accessed 24 July 2018
14. Oprunenco, A., Akmeemana, C.: Using blockchain to make land registry more reliable in India. In: LSE Business Review. The London School of Economics (2018). http://blogs.lse.ac.uk/businessreview/2018/04/13/using-blockchain-to-make-land-registry-more-reliable-in-india/. Accessed 23 July 2018

15. Pinna, A., Ruttenberg, W.: Distributed ledger technologies in securities post-trading. European Central Bank (2016). https://www.ecb.europa.eu/pub/pdf/scpops/ecbop172.en.pdf. Accessed 23 July 2018

16. Radzi, R.M., Lewis, M.K.: Religion and the clash of "Ideals" and "Realities" in business: the case of Islamic bonds (Sukuk). Thunderbird Int. Bus. Rev. **57**, 295–310 (2015)

17. Tapscott, A., Tapscott, D.: How blockchain is changing finance. Harv. Bus. Rev. **1** (2017)

18. Usmani, M.T.: An Introduction to Islamic Finance. Arham Shamsi (2000)

19. Vakta, T., Maheswari, A., Mohanan, N.U.: Blockchain disruption in security issuance. Capgemini (2016). https://www.capgemini.com/wp-content/uploads/2017/07/blockchain_securities_issuance_v6_web.pdf. Accessed 20 Sept 2018

20. Wall, E., Malm, G.: Using blockchain technology and smart contracts to create a distributed securities depository. In: Department of Electrical and Information Tehnology. Lund University, Lund, Sweden, p. 81 (2016)

21. Weiss, M., Corsi, E.: Bitfury: blockchain for government. HBS Case Study 818-031, 12 January 2017

22. Wood, G.: Ethereum: a secure decentralised generalised transaction ledger. Ethereum Project Yellow Paper 151 (2014). http://gavwood.com/paper.pdf. Accessed 25 July 2018

23. Zheng, Z., Xie, S., Dai, H., et al.: An overview of blockchain technology: architecture, consensus, and future trends. In: 2017 IEEE International Congress on Big Data (BigData Congress), pp. 557–564 (2017)

Market Data Analytics

The Role of Customer Retention in Business Outcomes of Online Service Providers

Behrang Assemi[1,2](✉) (iD)

[1] UNSW Business School, University of New South Wales (UNSW) Sydney,
Sydney, Australia
b.assemi@uq.edu.au
[2] Science and Engineering Faculty,
Queensland University of Technology (QUT), Brisbane, Australia

Abstract. Despite great business outcomes that some service providers achieve on crowdsourcing marketplaces, many are unable to find customers, transact on a regular basis and survive. Previous research on the customer–provider relationship development in these marketplaces has shown the impact of provider profiles on customers' decision-making and highlighted the role of prior relationship between the two parties as a major determinant of provider selection decisions of customers. However, little is known about the role of prior relationship and customer retention on the business outcomes of providers. This study theorizes a mediating impact versus a moderating impact of customer retention on the association between profile information and business outcomes of providers, building on the integrated information response model (IIRM). This study investigates two competing theoretical models using archival data from a leading crowdsourcing marketplace. The results show that a provider's profile information is significantly associated with the provider's business outcomes, while customer retention partially mediates this relationship for the self-provided information on the profile. Moreover, customer retention negatively moderates the positive impact of the customer-provided information on the provider's business outcomes, implying the important role of new customers in achieving better business outcomes on crowdsourcing marketplaces.

Keywords: Crowdsourcing marketplace · Online service provider ·
Business outcomes · Customer retention ·
Integrated information response model (IIRM) · Partial least squares (PLS)

1 Introduction

Despite great business outcomes achieved by some crowdsourcing marketplace providers, many are unable to find customers, transact on a regular basis and survive in these marketplaces [1]. As of 2016, three major, highly-visited crowdsourcing marketplaces [according to 2]—Upwork.com, Freelancer.com and Guru.com—had 33 million registered users, and a total transaction volume of US$3.7 billion [3–7]. According to most recent financial reports, about US$1.5 billion of services are transacted annually through these crowdsourcing marketplaces [8–10]. However, on Elance.com as an example, only nine percent of the registered providers have gained

© Springer Nature Switzerland AG 2019
N. Mehandjiev and B. Saadouni (Eds.): FinanceCom 2018, LNBIP 345, pp. 79–100, 2019.
https://doi.org/10.1007/978-3-030-19037-8_6

money, whereas 17 providers have earned more than US$400,000 during the year ending at January 2013 [11]. Such a high failure rate of online service providers is contrary to the survey findings indicating that these providers are primarily registering on crowdsourcing marketplaces for steady work and revenue [e.g., 12, 13].

Given a lack of real-world acquaintance between most customers and providers in a global crowdsourcing marketplace, providers' public profiles are the only sources of information for customers in the early stages of relationship development between the two parties [14–17]. The information on these profiles signals the reputation and past performance of their owners, and thus impacts on customers' decisions as to whom to outsource their service projects [14, 18, 19]. Customers' decisions and transacting behavior in turn, affect providers' business outcomes [1].

The literature on crowdsourcing marketplaces has investigated the impact of some common information components on provider profiles on the provider selection decisions of customers [e.g., 14, 16, 20], auction outcomes [e.g., 21, 22] and project pricing [e.g., 23]. However, the impact of providers' profile information on customer retention, and the role of customer retention in business outcomes of providers are yet underexplored.

A higher customer retention rate can positively impact on the business outcomes of suppliers (providers) in terms of their revenue and growth in electronic marketplaces in general [24–26], by replacing the costs of new customer acquisition with the considerably lower expenses of maintaining existing customers [25–27]. However, the findings of literature about the impact of prior relationship between customers and providers on the provider selection decisions of customers, and thus the business outcomes of providers are fragmented. Moreover, the competitive market-oriented structure of crowdsourcing marketplaces, as anticipated by Malone et al. [28], leads customers to arm's length relationships rather than long-term repeat transactions with providers [29, 30]. This raises a strong demand for a solid investigation of the impact of customer retention on business outcomes of providers on crowdsourcing marketplaces. Accordingly, this study aims to address such a need by seeking the answer to the following research questions:

- Does customer retention mediate the relationship between profile information and business outcomes of providers on crowdsourcing marketplaces?
- Does customer retention moderate the relationship between profile information and business outcomes of providers on crowdsourcing marketplaces?

Therefore, two theoretical models are developed based on the integrated information response model (IIRM) [31] to evaluate the potential mediation or moderation impact of customer retention on the business outcomes of providers in crowdsourcing marketplaces. The models are empirically tested through partial least square (PLS) analysis, using the archival profile information of service providers collected from a leading crowdsourcing marketplace.

The remainder of this document is structured as follows. The next section summarizes the key findings of the literature on the business outcomes of crowdsourcing marketplace providers and the potential impact of customer retention on these outcomes. Then, the two alternative theoretical models are presented. Next, the research method is discussed. The data analysis and results are presented next. Then, the findings of this research are presented and the implications of this research for both theory and practice are discussed. Finally, some concluding remarks are provided in the last section.

2 Literature Review

The major business concern of online providers is to increase their revenue from both new and existing customers [32]. The success of providers in achieving this objective can significantly influence the participation of both providers and customers, which in turn impacts on the success of the entire marketplace [33]. However, the literature on crowdsourcing has mainly investigated the impact of some profile information components on the provider selection decision of customers, auction outcomes and project pricing. Thus, the impact of information presented on crowdsourcing marketplace provider profiles on customer retention and the business outcomes of providers have remained underexplored.

Holthaus and Stock [18] found significant, positive associations between the following components on a provider's profile and their annual earnings on a crowdsourcing marketplace: experience (in terms of the number of past projects), English proficiency, number of portfolio items, average rating, number of passed skill assessments and hourly rate. They also found significant associations between ingratiation toward customers and self-promotion in a provider profile's self-description and the provider's annual earnings, while the latter is much stronger compared to the former. Holthaus and Stock [18] also showed that being from countries that are known as major outsourcing destinations (i.e., India and Asian countries) is positively associated with a provider's annual earnings. However, they did not find a significant association between a provider's education and their annual earnings.

Assemi and Schlagwein [17] evaluated the impact of some common information components presented on provider profiles on the business outcomes of providers in terms of their annual revenue. They classified these information components into two major categories relying on the information source: the information published by providers themselves and the information which reflects previous customers' feedback. Assemi and Schlagwein [17] showed that the former significantly affects the business outcomes of providers, whereas the latter does not. In the first category, the number of sample job items and the number of skills assessed as in the top 10% of providers are the information components which significantly contribute to their corresponding exogenous construct.

Most previous studies, however, have focused on evaluating the impact of information components presented on provider profiles, on the provider selection decisions of customers. Average (weighted) rating is the most studied information component on provider profiles which has been shown that significantly influences the provider selection decisions of customers [14, 16, 22]. Gefen and Carmel [15] as well as Hong and Pavlou [20] showed that the country of origin of providers also influences the decision making of customers. Gefen and Carmel [15] found that customers often prefer to outsource their services to providers from their own country of origin. The findings of Kim [30] as well as Gefen and Carmel [16] confirmed this result. Gefen and

Carmel [15] also showed that whenever customers decide to outsource their projects to offshore destinations, they generally prefer to exploit the labor arbitrage by outsourcing their projects to providers from countries with inferior purchase power parities. Moreover, customers from English-speaking countries often favor English-speaking providers [15]. Kim [30] as well as Hong and Pavlou [20] showed that a provider's experience (in terms of the number of completed projects) has also a positive impact, while Kim [30] found that a provider's firm size (in terms of the number of employees) has a negative impact on the provider selection decisions of customers. Banker and Hwang [14] illustrated that a provider's greater total earnings, reputation status, standardized service offering and membership in the dispute resolution program of the marketplace positively influence the provider selection decisions of customers. Finally, Banker, Wattal [1] showed that a higher duration of professional experience, number of references, average weighted rating, and marketplace preference status of a provider are positively associated with the decisions of customers to transact with the provider.

The literature has also investigated the impact of providers' profile information on project pricing and auction outcomes. Gefen and Carmel [34] found that a higher average rating of a provider and a larger number of their feedback ratings positively influence the complete payment of the provider's projects by the corresponding customers when the provider's average rating is above the stated average of the crowdsourcing marketplace. Hong and Pavlou [23] showed that the type (business vs. individual), purchasing power (in terms of purchase power parity of the country of origin), average rating, experience (in terms of total earnings), hourly wage and auction success rate of a provider are significantly associated with the bid prices proposed by the provider, "price premium[s]" and "bid-ask spread[s]". They characterized the price premium as the difference between the price proposed by the provider and the average price proposed by all providers who bided in the auction divided by the average bid price in the auction. They also defined the bid-ask spread as the difference between the price proposed by the provider and the budget specified by the customer for a project divided by the specified budget. Hong and Pavlou [23] also illustrated that prior satisfactory contracts between a provider and a customer negatively affects the provider's proposed bid price for that customer, while a better marketplace reputation status of a provider is positively associated with their price premiums and bid-ask spreads.

Finally, the literature has shown that prior relationship between a customer and a provider significantly influences the customer's decision to outsource their projects to the provider. Banker and Hwang [14] as well as Hong and Pavlou [20] found that inviting a provider by a customer to bid in an auction positively influences the customer's decision to outsource the corresponding project to the provider. Radkevitch, van Heck [35] showed that there are categories of customers on crowdsourcing marketplaces that often favor transacting with their preferred providers. Kim and Wulf [29, 36] illustrated that customers prefer to repeatedly transact with a small number of providers instead of exploring new ones when they become more experienced in a crowdsourcing marketplace. Gefen and Carmel [15, 16] demonstrated the significant positive impact of prior relationship between a customer and a provider on the customer's decision to transact with the provider, although Kim [30] found such an impact to be insignificant. Gefen and Carmel [16] also showed that a higher number of transactions between the two parties positively influences the customer's decision to outsource their new projects to the provider. Finally, previous research found that customers who repeatedly transact with their preferred providers often possess greater award ratio [29, 35] and total transaction value [35].

However, the impact of customer retention on the business outcomes of providers is overlooked in the literature. Although previous research has shown that some providers prefer to repeatedly transact with their preferred providers, little is known about the impact of such repeat transactions on the business outcomes of providers. Furthermore, while a significant association between providers' profile information and the provider selection decisions of customers has been found in the literature, the potential impact of customer retention on this association is not explored well. To fill this gap, this research investigates the role of customer retention in the business outcomes of providers in more detail.

3 Theoretical Background and Model Development

3.1 Integrated Information Response Model

This study's theoretical models are built on the IIRM proposed by Smith and Swinyard [31]. According to Smith and Swinyard [31], a customer's exposure to information about a supplier (provider) determines the sequence of their behavioral responses to the supplier, as shown in Fig. 1. Such a sequence is always initiated by the customer's 'cognition' about the supplier which is formed relying on the available information. This cognition is associated with the customer's emerging 'beliefs' about the supplier, namely the likelihood of attributing specific characteristics (e.g., trustworthiness) to the supplier. When the information is perceived as of low credibility by the customer, it can form 'lower order beliefs' which potentially persuade the customer to collect more information about the supplier through a trial 'conation' (e.g., direct negotiation or a trial exchange transaction with the supplier). Such a trial conation may further form the customer's higher order beliefs and attitudes toward the supplier. When the information is perceived as of high credibility by the customer, it can result in 'higher order beliefs' which affect the customer by forming their higher order 'liking', 'preference' and 'conviction' attitudes. Such attitudes in turn, can lead the customer to commitment toward the supplier.

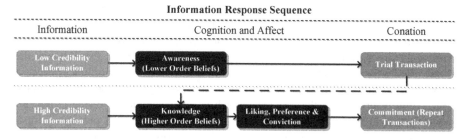

Fig. 1. The IIRM, adapted from Smith and Swinyard [31]

Relying on the IIRM, this study argues that providers' profile information can significantly impact on the business outcomes of the respective providers. On the one hand, the information components presented on these profiles can persuade customers to conduct trial exchange transactions with the corresponding providers when the customers perceive the information as of low credibility. For example, the profile information which is published by providers about themselves is potentially considered as of lower credibility because such information is suspected to be biased by the information sources' self-interest [31, 37]. On the other hand, the information components on provider profiles can cause customers to repeatedly transact with the same providers when they perceive the respective information as of higher credibility. For example, the information which is verified by a neutral third party is potentially considered as of higher credibility [31, 37, 38].

Accordingly, two alternative theoretical models (illustrated in Figs. 2 and 3) are proposed in this study. Both models classify information presented on a provider profile into two distinct categories based on the source of information: self-provided information and customer-provided information. Self-provided information encompasses the information which is published by providers about themselves. Thus, the respective information components are potentially perceived as of lower credibility, especially given the potential likelihood of information bias caused by the information source's self-interest [31]. Customer-provided information entails the information components which reflect the feedback ratings and recommendations posted by previous customers about a given provider. Such information is possibly considered as of higher credibility, especially because of the rather neutral source of information [31]. The first model, namely model A (shown in Fig. 2), theorizes the potential impact of the two categories of information on provider profiles on the business outcomes of the respective providers, both directly and through enhancing customer retention toward the providers (i.e., customer retention is incorporated in the model as a mediator). Model B (shown in Fig. 3) is proposed based on the findings of the literature to evaluate a potential moderating impact of customer retention on the hypothetical associations between the two categories of profile information and the business outcomes of the corresponding providers. These two theoretical models are explained next.

3.2 Hypothesis Development: Customer Retention as a Mediator

Building on the IIRM, the self-provided information on a provider profile is potentially perceived as of low credibility by customers, and thus can persuade them to conduct trial exchange transactions with the provider to collect more information. Such trial transactions directly influence the business outcomes of the provider. Furthermore, the self-provided information indicates the extent of the provider's trustworthiness because the respective information components imply the possession of specific technical abilities and experience by the provider, which is a major indicator of the provider's trustworthiness [14, 16, 23, 37, 39]. According to the literature, a provider's trustworthiness can directly impact on the provider's business outcomes [34, 40–42] because it decreases the perceived risks and uncertainty related to transacting with the provider [25, 43, 44] and encourages customers to transact with the provider [26, 44]. Therefore, this study hypothesizes:

Hypothesis 1a: The more affirmative the self-provided information is on a provider profile, the more likely is the provider to obtain large business outcomes.

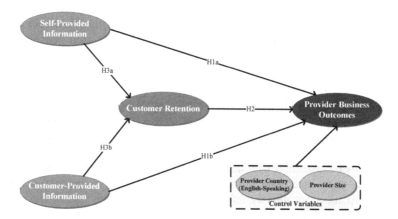

Fig. 2. Theoretical model A: customer retention as a mediator

By contrast, the customer-provided information on a provider profile is potentially perceived as of high credibility by customers, and thus can cause the existing customers to repeatedly transact with the provider. Such repeat transactions also directly influence the business outcomes of the provider. Furthermore, customer-provided information indicates previous customers' satisfaction with a given provider and the provider's overall reputation on the marketplace [43, 45]. Previous customers' satisfaction signals the reputation and trustworthiness of the provider to new customers [43, 46], which are major determinants of repeat transactions with the provider [33, 34, 47, 48]. A provider's reputation is an important indicator of their trustworthiness because a good reputation significantly decreases the risks of transacting with the provider as perceived by customers [39, 42, 43, 49]. Both customers' satisfaction with a provider and the provider's trustworthiness are major determinants of the attitudes of customers toward the provider and directly impact on the customers' intentions to transact with the provider [33, 34, 43, 47, 48]. Therefore, this study hypothesizes:

Hypothesis 1b: The more affirmative the customer-provided information is on a provider profile, the more likely is the provider to obtain large business outcomes.

A higher customer retention rate can positively influence the provider's business outcomes, and thus potentially mediates the relationship between the provider's profile information and the provider's business outcomes. The customer retention rate is directly associated with the 'certainty' and 'growth' of the provider's exchange relationships with customers [24]. Customer retention reduces the costs of transactions for the provider by substituting new customer acquisition's costs with considerably lower expenses of maintaining the existing customers [25–27]. Furthermore, retained customers are more flexible regarding the provider's faults, less sensitive to proposed prices [24, 25], and more willing to repeat transactions with the provider [15]. Therefore, this study hypothesizes:

Hypothesis 2: A higher customer retention rate of a provider positively impacts on the provider's business outcomes.

The self-provided information on a provider profile can also lead customers to commitment toward the provider, as it indicates the level of competence and trustworthiness of the provider [14, 16, 23, 37, 39]. As discussed earlier, customers consider transacting with a more trustworthy provider to be less risky, and hence are more willing to transact with such a provider [34, 40, 43, 44]. Therefore, this study hypothesizes:

Hypothesis 3a: The more affirmative the self-provided information is on a provider profile, the more likely is the provider to have a higher customer retention rate.

The customer-provided information on a provider profile is potentially considered as of high credibility by customers, and thus can lead customers to commitment toward the provider, according to the IIRM. Moreover, the relevant information indicates the level of customer satisfaction with the provider [25, 27, 50]. Customer satisfaction is also positively associated with customer retention [44, 50]. Therefore, customer-provided information can directly affect the customer retention rate of the provider. Consequently, this study hypothesizes:

Hypothesis 3b: The more affirmative the customer-provided information is on a provider profile, the more likely is the provider to have a higher customer retention rate.

Figure 2 illustrates the proposed hypotheses (i.e., model A). As shown in the figure, two characteristics of service providers (i.e., provider country and size) that are shown by the literature to impact on customers' choices are also included in the analyses as control variables. As discussed in Sect. 2, providers from English-speaking countries and those with smaller firm sizes potentially obtain greater business outcomes.

3.3 Hypothesis Development: Customer Retention as a Moderator

Customer retention, however, can enhance the potential impact of a provider's profile information on the business outcomes of the provider. The IIRM suggests that a committed customer reacts to information about a given provider differently compared to a new customer [31]. Previous research has also showed that the former considers less information in their decision making while transacting with their preferred providers [51]. As discussed in Sect. 2, there are categories of customers on crowd-sourcing marketplaces who prefer to repeatedly transact with a small group of providers [30, 35, 36]. Therefore, this study hypothesizes:

Hypothesis 4: A higher customer retention rate of a provider positively impacts on the association between (a) the self-provided information as well as (b) the customer-provided information on the provider's profile and the provider's business outcomes.

Figure 3 demonstrates these hypotheses (i.e., model B), in which customer retention moderates the relationship between a provider's profile information and the provider' business outcomes.

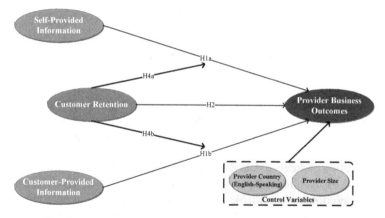

Fig. 3. Theoretical model B: customer retention as a moderator

4   Research Method

4.1   Research Context

Crowdsourcing marketplaces are global virtual exchange environments through which customers (also known as requesters, employers or clients) commercially outsource their required services to providers (also known as freelancers, suppliers or crowdworkers) who possess the needed skills and qualities [23, 52, 53]. The data for this research were collected from a leading crowdsourcing marketplace. Service projects outsourced in this marketplace encompass a wide variety of services, such as software development, financial planning, marketing and virtual administration. Only in 2012, almost 800,000 new providers registered in the marketplace to transact over 800,000 service projects requested by customers and they earned approximately US$200 million in total.

The sequence of customer–provider interactions in a typical exchange transaction in such a crowdsourcing marketplace is usually initiated by a customer. The customer creates an online auction by posting a request for proposal for the intended services. Relying on the profile information of providers and prior exchange relationships with them, the customer may proactively invite some providers to participate in the auction. The customer can even restrict the auction participants to the invited providers. Subsequently, providers bid in the auction by proposing their technical solution, price and estimated delivery time. When the auction is closed, the customer awards the project to a service provider who is perceived as to be the best fit for the project. The information components on provider profiles (as the major source of information available to customers about providers) play a significant role in determining customers' decisions to (repeatedly) transact with providers, which in turn can affect the providers' business outcomes.

4.2   Data Collection

The data were collected from a leading, well-known crowdsourcing marketplace in 2013 and entail the profile details of all "IT (Information Technology) and Programming" firms who have earned at least US$1,000 during a year leading to the time of the data collection. The values for all the measures in the research models were extracted from the provider profiles for the period of 12 months preceding the data collection. The original data set comprised 831 provider profiles. Four extreme cases with annual earnings more than US$1 million were excluded and 827 provider profiles remained for the analysis.

The application of archival data is associated with several advantages for this study. As emphasized by Tomlinson-Keasey [54], archival data sets provide comprehensive and accurate descriptions of variables, improve predictions relying on a large number of records, enable evaluation of moderating/mediating effects, determine the changes in variables and their interrelationships in a longitudinal manner and provide a clearer image of causal directions between variables. Furthermore, having relied on the archival data in this research, the measures were associated with small or no assumed measurement errors because their values were directly observed on the crowdsourcing marketplace website.

The sample collected for this research fairly represents the profiles of other types of service providers in a large crowdsourcing marketplace. The IT and software development categories often encompass the highest level of user participation in well-known crowdsourcing marketplaces [21, 55]. Moreover, inherent intangibility and customizability of software development services make them reasonable representatives of other types of services [29, 36]. Finally, a high level of similarity among the structure of provider profiles and market mechanisms in well-known crowdsourcing marketplaces means that the collected data reasonably represent the patterns of data in other crowdsourcing marketplaces, although any generalization of this study's results to other crowdsourcing marketplaces requires further investigation.

4.3   Measurement

Dependent Variable and Mediator/Moderator. The dependent variable in both models, namely the business outcomes of a provider, was measured by the annual revenue of the provider in the crowdsourcing marketplace. Measuring a provider's business outcomes through this variable on crowdsourcing marketplaces is in line with previous studies in the literature. A provider's outcomes can be measured at different levels of analysis [56]. The business strategy level is the highest level of analysis, in which metrics such as earnings, net income, profit, compound total return, return on investment, number of contracts and number of customers have been used by previous research to measure the outcomes of a provider [37, 56–58]. For crowdsourcing marketplaces, Holthaus and Stock [37] have proposed the number of contracts and total earnings over a period of time to objectively measure the business outcomes of providers. Similarly, Holthaus and Stock [18] have used annual earnings as the main measure of the business performance of crowdsourcing marketplace providers.

Accordingly, this study has used a provider's revenue, total customers and total con-tracts over one year to measure the dependent variable.

The mediator/moderator construct of this research, namely customer retention, was measured through the share of repeat transactions of a given provider with their existing customers over one year. This indicator has been previously adopted by many studies to operationalize customer retention [e.g., 26, 59].

Independent Variables. As discussed before, the information components presented on a provider profile can be classified into two major categories, given the source of information: self-provided information and customer-provided information. Self-provided information encompasses the information which is published and updated by providers themselves. Across well-known crowdsourcing marketplaces, the com-mon self-provided information components on a provider profile include: the provi-der's experience on the crowdsourcing marketplace (i.e., the number of months passed since their registration), number of portfolio items, number of certificates, number of skills assessed as in the top 10% of providers on the marketplace and total number of credentials (e.g., references) verified by the crowdsourcing marketplace. Customer-provided information entails the feedback ratings and recommendations submitted by the customers who have transacted with the provider. The common customer-provided information components on a provider profile in well-known crowdsourcing market-places consist of: the average customer recommendation, average weighted rating across all projects (weighted by the corresponding project values), and numerical ratings for individual projects. The indicators of both constructs (i.e., self-provided information and customer-provided information) are 'formative' [60, 61], because: (1) their combination forms the corresponding latent construct, (2) they are not cor-related, and (3) they do not always vary in the same way when the respective construct is altered.

Control Variables. In this study's theoretical models, two control variables, namely provider language and size, were incorporated. As discussed by Gefen and Carmel [15], customers are more likely to crowdsource their services to offshore providers when both parties are from English-speaking countries. Hence, providers from English-speaking countries may obtain better business outcomes. Provider language was con-sidered in the models using a categorical, binary indicator which is one for English-speaking providers and zero for others. Furthermore, as discussed in the literature, customers often prefer providers with smaller firm sizes [30]. Thus, given the range of each provider's firm size specified on their profiles, the mean of the range was used to measure the provider size.

4.4 Data Analysis

Partial least squares (PLS) was used to analyze the data. PLS has gained great popu-larity in information systems (IS) research, mainly because of its ability to assess complex structural models including reflective and/or formative constructs, relying on small- to medium-sized samples with any type of distribution [61, 62]. It is a second-generation multivariate technique that allows researchers to specify and simultaneously estimate relationships between multiple constructs as well as the indicators of each construct in a single theoretical model [62–64]. Therefore, PLS suits the analysis of this research, especially because of non-normal formative measures as well as mediation/moderation relationships incorporated in the research models.

SmartPLS Version 3.0 [65] was used in this study to evaluate the proposed hypotheses. The sample size of 827 was enough for the analysis of both models, because the minimum sample size required for a PLS analysis is 10 times the largest number of structural paths heading to a construct in the model [60], namely 80 for model A and 70 for model B. According to Hair et al. [64], multicollinearity was not an important concern for the analysis of model A or model B, because the variance inflation factors (VIFs) were all less than 10 in both models. A PLS bootstrapping test [61] with 1,000 runs was conducted to evaluate the significance of relationships.

To evaluate the moderation effect of customer retention in model B, the two-stage approach proposed by Chin et al. [66] and further discussed by Henseler and Fassott [67] as well as Henseler and Chin [68] was applied. According to Henseler and Chin [68], a moderation effect cannot be evaluated using the pairwise product of indicators (of the corresponding constructs) when the moderator variable and/or the exogenous variable are formative. Instead, a two-step PLS method is suggested by Chin et al. [66], in which the scores of the latent constructs are estimated first in the main effect model [67, 68]. In the second step, the standardized scores of the exogenous and moderator variables are multiplied to create the interaction term, which is then applied along with the standardized scores of the relevant constructs in a multiple linear regression analysis [67]. In this multiple regression analysis, the scores of the endogenous variables are used for the dependent variables [67, 68].

5 Results

Table 1 illustrates the descriptive statistics and variance inflation factors (VIFs) of the indicators in both theoretical models. As shown in Table 1, all VIFs are less than 10, which indicates that multicollinearity does not cause a detrimental effect [64]. Measurement validity and reliability are not relevant concerns for this research, because the indicators of both independent constructs, namely the self-provided information and the customer-provided information, are formative [60, 69, 70]. According to Neuman [69], measurement reliability is often a challenge for the studies which incorporate complex latent constructs that are measured mainly relying on the perception of research participants. However, as described in the data collection section, this study's indicators were directly observed in the archival data with no assumed measurement errors. The applied data collection technique also avoids any common method variance [71, 72].

The analysis results are discussed in the remainder of this section to demonstrate the role of customer retention in business outcomes of service providers. First, the results of the analysis of model A are presented, in which customer retention is a mediator between the provider profile information and the provider business outcomes. Then, the results of the analysis of model B are presented, in which customer retention moderates the relationship between the provider profile information and the business outcomes. Finally, the results of a hierarchical difference test are presented which was conducted to compare the explanatory power of these two models.

Table 1. Descriptive statistics (N = 827)

Construct	Measure	Range	Mean	Std. Dev.	VIF
Provider business outcomes (Dependent variable)	Annual Customers	0–114	8.0	11.6	5.94
	Annual Earnings	1000–607653	19982.3	48714.2	1.27
	Annual Contracts	0–126	11.4	16.6	5.91
Self-provided information (Independent variable)	Verified Credentials	0–14	0.35	1.19	1.53
	Skills Assessed as in Top 10%	0–14	0.43	1.32	1.60
	References	0–14	0.61	1.54	1.38
	Group Memberships	0–22	1.33	2.30	1.77
	Top Team Members	0–4	1.20	1.55	1.25
	Portfolio Items	0–448	19.64	44.36	1.12
	Certificates	0–27	0.58	1.86	1.22
Customer-provided information (Independent variable)	Recommendation	0–100	42.40	48.06	1.34
	Average Rating	0–5.0	3.99	1.80	1.22
	Average Weighted Rating 2 Weeks	0–5.0	1.48	2.22	1.24
Customer retention (Mediator or moderator)	Customer Retention Rate	0–100	29.37	31.25	1

5.1 Customer Retention as a Mediator

Figure 4 shows the results of PLS bootstrapping analysis for model A (incorporating customer retention as a mediator). The model explains 49.9% of the variance in provider business outcomes. The model also accounts for 1% of the variance in customer retention. As hypothesized, the path coefficients for the association between self-provided information and provider business outcomes (H1a, $\beta = 0.406$), as well as the association between customer-provided information and provider business outcomes (H1b, $\beta = 0.415$) are both positive and significant at $\alpha = 0.001$. However, the association between customer retention and provider business outcomes (H2) is significant at $\alpha = 0.1$, but negative ($\beta = -0.035$). Finally, the path coefficient for the association between self-provided information and customer retention (H3a, $\beta = 0.057$) is positive and significant at $\alpha = 0.05$, while the path coefficient for the association between customer-provided information and provider business outcomes (H3b, $\beta = 0.050$) is not significant. Thus, customer retention negatively mediates the relationship between self-provided information and provider business outcomes, although the corresponding direct impact is larger.

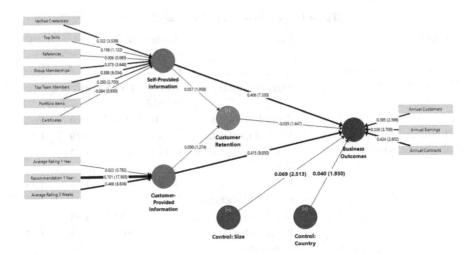

Fig. 4. PLS results for model A (Customer retention as a mediator)

As shown in Fig. 4, among the indicators of self-provided information, four contribute significantly to the corresponding construct: verified credentials, group memberships, top team members and portfolio items. Similarly, among the indicators of customer-provided information, two contribute significantly to the respective construct: recommendation over one year and average weighted individual rating over two weeks. Finally, the three indicators of provider business outcomes are all significantly associated with the corresponding construct.

The results of PLS bootstrapping analysis for model B (incorporating customer retention as a moderator) are illustrated in Fig. 5. The model accounts for 50.7% of the variance in provider business outcomes. The path coefficients for self-provided information ($\beta = 0.393$) and customer-provided information ($\beta = 0.421$) are both significant at $\alpha = 0.001$. The association between customer retention and provider business outcomes is not significant anymore. Customer retention significantly moderates the association between customer-provided information and provider business outcomes (H4b, $\beta = -0.100$) at $\alpha = 0.001$, although it suppresses the impact of customer-provided information. Finally, customer retention does not significantly moderate the association between self-provided information and provider business outcomes.

5.2 Model Comparison

To compare model B (i.e., the model with moderation effects) with model A (i.e., the model with mediation effects), a "hierarchical difference test" was conducted, as recommended by Chin et al. [66] and applied by previous research [e.g., 70]. Accordingly, the difference between the R-squares of model B and model A was used to evaluate the effect sizes of the moderation effects. As illustrated in Table 2, the hierarchical difference test shows that the effect size f^2 of the moderation effects in model B is 0.02, which is considered as a small effect. While model A accounts for 49.9% of the

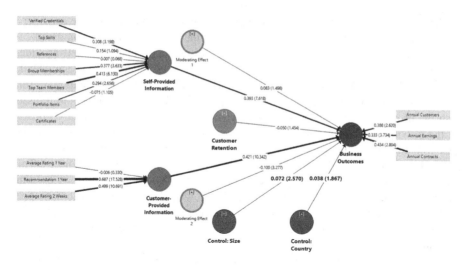

Fig. 5. Two-step PLS results for model B (Customer retention as a moderator)

variance in provider business outcomes, the inclusion of the two moderation effects in model B increases the variance explained to 50.7% (Figs. 4 and 5). As illustrated in Fig. 5, customer retention significantly moderates the association between customer-provided information and provider business outcomes. Thus, model B in which customer retention is proposed to moderate the relationships between self-provided information as well as customer-provided information, and provider business outcomes has a slightly higher explanatory power.

Table 2. Model comparison

Model	R^2	f^2
A (Customer retention as a mediator)	0.499	0.02
B (Customer retention as a moderator)	0.507	

The effect size f^2 is calculated as [R^2(model with moderation effects) – R^2(model without moderation effects)]/[1 – R^2(model without moderation effects)] [66]. The effect size f^2 larger than 0.02, 0.15 and 0.35 are respectively considered as small, medium and high [73].

6 Discussion

Two theoretical models were developed and empirically evaluated in this study based on the IIRM [31] to investigate the role of customer retention in the business outcomes of providers. In one model (model A), customer retention was incorporated as a mediator for the association between a provider's profile information and the provider's business outcomes. In another model (model B), customer retention was included to moderate the relationship between a provider's profile information and the provider's business outcomes. The results of the analysis of the two models were compared.

The analysis of model A (shown in Fig. 4) indicates that both self-provided information and customer-provided information on a provider profile significantly impact on the provider's business outcomes. This finding confirms the important role of a provider's profile information in attracting (new and repeat) customers to transact with the provider. A provider's customer retention rate, however, negatively impacts on the provider's business outcomes. This surprising finding indicates that the providers who transact more with new customers rather than their existing customers are more likely to obtain higher business outcomes on crowdsourcing marketplaces. This can be attributed to the fact that providers often propose lower prices to their existing customers [23]. Moreover, when transacting more with their existing customers, providers miss the opportunity to expand their customer-base and professional experience on the crowdsourcing marketplace, which in turn negatively impacts on their business outcomes.

The results of analysis also indicate that customer retention partially mediates the relationship between the self-provided information on a provider profile and the provider's business outcomes. The results show that a higher customer retention rate is more likely when a provider has more affirmative self-provided information on their profile. Given the significant, negative impact of the customer retention rate on the provider's business outcomes, however, these results indicate that a higher customer retention rate decreases the effect of self-provided information. This is mainly because existing customers are less likely to rely on the provider's profile information for repeatedly transacting with the provider.

The analysis of model B (illustrated in Fig. 5) indicates that customer retention only moderates the association between customer-provided information and provider business outcomes. Customer retention, however, suppresses the impact of customer-provided information on provider business outcomes, rather than enhancing it. This indicates that having a larger ratio of repeat customers limits the business development opportunities for the corresponding provider and negatively impacts on their business outcomes.

The comparison of the two models (as presented in Table 2) indicates that model B, in which the customer retention is incorporated as a moderator, has a higher explanatory power compared to model A. Although the difference between the two models' explained variance is small, model B is preferred, given the small, partial mediation effect of customer retention found in model A.

6.1 Implications for Theory

The results of this research indicate that the behavioral responses of customers to providers' profile information follow the sequence anticipated by the IIRM [31]. As illustrated above, the self-provided information on provider profiles, which is potentially perceived as of lower credibility by customers and leads customers to trial exchange transactions with providers, positively impacts on providers' business outcomes. Similarly, the customer-provided information on provider profiles, which is possibly perceived as of higher credibility by customers and causes customer commitment toward providers, positively impacts on providers' business outcomes as well. However, in crowdsourcing marketplaces, arm's length relationships are more salient and customer commitment and retention do not have a large, positive impact on the business outcomes of providers.

This research is among the first studies which have investigated the impact of providers' profile information on the providers' business outcomes in crowdsourcing marketplaces. Provider profiles represent one of the most prominent differences between crowdsourcing marketplaces and conventional service outsourcing markets, as these profiles provide customers with a comprehensive access to information about the past transactions and contextual behavior of the providers in the crowdsourcing marketplace. This study's theoretical models are useful in both explaining and forecasting the business outcomes of crowdsourcing marketplace providers based on their profiles' information as well as customer retention rate.

6.2   Implications for Practice

Although previous research has highlighted the significant impact of prior relationship between a customer and a provider on the customer's decision to outsource their new projects to the provider [15, 16], the findings of this study show that customer retention does not positively influence providers' business outcomes. This means that customers prefer arm's length relationships which maximizes their benefit in the short-term. Thus, although online service providers need to maintain their relationships with the existing customers at a satisfactory level, especially to obtain better feedback, they cannot rely on repeat transactions with their existing customers to obtain better business outcomes. Instead, they are required to invest in improving and updating their self-provided information to better illustrate their trustworthiness and acquire more new customers.

These results confirm the previous literature findings that exploiting the power of crowd can disrupt the traditional markets in the future; however, it currently complements the existing markets by providing transacting parties with novel exchange alternatives [74]. While there are a wide range of motivations for both customers and providers to participate in a crowdsourcing marketplace, immediate/short-term financial return still plays an important role in encouraging marketplace participation [74]. In contrast to conventional markets, arm's length relationships through which both customers and providers can explore new talents to further benefit from an easy access to a large, readily available skilled labor pool [75] is more prevalent on crowdsourcing marketplaces [29, 30]. Accordingly, these marketplaces should thoroughly support such relationships to warrant active participation of both parties, which is necessary for the success of the marketplace [33, 76, 77].

6.3   Limitations and Outlook for Future Research

This research has several limitations which can lead to opportunities for future research.

First, the archival data of provider profiles were collected only for "IT and Programming" provider firms. Although the sample was chosen to be representative of other providers on the crowdsourcing marketplace, the results may not be generalizable to providers working in other service categories or in other crowdsourcing market-places. Consequently, further research is required to evaluate the proposed research models across different service categories and crowdsourcing marketplaces.

Second, this study has mainly aimed at contrasting the mediating versus moderating effect of customer retention on the association between profile information and business outcomes of providers in a crowdsourcing marketplace. Indeed, no causal relationship was theorized and investigated in the theoretical models. However, a longitudinal analysis can be done with two datasets collected at different points in time for the same providers to shed more light on potential causality between the theoretical constructs of interest.

Third, this study has focused on the impact of profile information on customer retention and provider business outcomes in crowdsourcing marketplaces. However, other factors such as providers' bidding behavior (including pricing) and auction characteristics may also influence customer retention and/or business outcomes of the providers [14, 16, 23]. Therefore, it is necessary to further evaluate such effects to reveal different determinants of customer retention and provider business outcomes in crowdsourcing marketplaces.

7   Conclusion

This study proposed and empirically evaluated two theoretical models incorporating customer retention as a mediator or moderator for the association between profile information and provider business outcomes on crowdsourcing marketplaces. Both models were built on the IIRM and were evaluated using empirical data collected from a leading crowdsourcing marketplace. The findings of this study contribute to our knowledge about the impact of profile information on the relationship development and business outcomes of providers in crowdsourcing marketplaces. The study results indicate that a provider's profile information significantly impact on the provider's business outcomes. Moreover, customer retention partially mediates the relationship between the self-provided profile information and provider business outcomes, while it moderates the association between the customer-provided profile information and provider business outcomes. To conclude, the findings of this study suggest that providers need to invest more on their profiles and consistently develop relationships with new customers (instead of only relying on existing customers) to obtain better business outcomes in a crowdsourcing marketplace.

References

1. Banker, R., Wattal, S., Hwang, I.: Determinants of firm survival in e-Markets: an analysis with software service providers. In: Hawaii International Conference on System Sciences (HICSS), Hawaii, USA (2011)
2. Alexa. Alexa Site Information (2016). http://www.alexa.com/. Accessed 01 Feb 2016
3. Freelancer.com. Full Year Results 2014 (2015). http://www.asx.com.au/asxpdf/20150216/pdf/42wm4m1b1r79z0.pdf. Accessed 13 Jan 2016

4. Freelancer.com. About the Company (2016). https://www.freelancer.com/. Accessed 13 Jan 2016
5. Guru.com. About the Company (2015). http://www.guru.com/. Accessed 13 Jan 2016
6. Upwork.com. Online Work Report: Global, 2014 Full Year Data (2015). http://elance-odesk.com/online-work-report-global. Accessed 13 Jan 2016
7. Upwork.com. About the Company (2015). https://www.upwork.com/about/. Accessed 13 Jan 2016
8. Upwork.com. About the Company (2018). https://www.upwork.com/about/. Accessed 26 Feb 2018
9. Barrie, M.: Freelancer 1H17 Half Year Results (2017). www.Freelancer.com
10. Guru.com. About the Company (2018). http://www.guru.com/. Accessed 26 Feb 2018
11. Elance.com. Elance Contractors (2013). https://www.elance.com/r/contractors/s-earningsSort/o-1/. Accessed 30 Jan 2013
12. Gandia, E.: Freelance Industry Report: Data and Analysis of Freelancer Demographics, Earnings, Habits and Attitudes. International Freelancers Academy (2011)
13. Elance.com. Elance Freelance Talent Report: A Look at the Demographics, Satisfaction Levels and Expectations of Online Freelancers (2010). https://www.elance.com/p/freelance-talent-report.html. Accessed 16 Apr 2012
14. Banker, R., Hwang, I.: Importance of measures of past performance: empirical evidence an quality of e-service providers. Contemp. Account. Res. **25**(2), 307–337 (2008)
15. Gefen, D., Carmel, E.: Is the world really flat? A look at offshoring in an online programming marketplace. MIS Q. **32**(2), 367–384 (2008)
16. Gefen, D., Carmel, E.: Why the first provider takes it all: the consequences of a low trust culture on pricing and ratings in online sourcing markets. Eur. J. Inf. Syst. **22**, 604–618 (2013)
17. Assemi, B., Schlagwein, D.: Profile information and business outcomes of providers in electronic service marketplaces: an empirical investigation. In: 23rd Australasian Conference on Information Systems. Deakin University, Geelong, Australia (2012)
18. Holthaus, C., Stock, R.M.: Facts vs. stories-assessment and conventional signals as predictors of freelancers' performance in online labor markets. In: Proceedings of the 51st Hawaii International Conference on System Sciences, Hawaii, USA (2018)
19. Assemi, B., Schlagwein, D.: Provider feedback information and customer choice decisions on crowdsourcing marketplaces: evidence from two discrete choice experiments. Decis. Support Syst. **82**, 1–11 (2016)
20. Hong, Y., Pavlou, P.A.: On buyer selection of service providers in online outsourcing platforms for IT services. Inf. Syst. Res. **28**(3), 547–562 (2017)
21. Snir, E.M., Hitt, L.M.: Costly bidding in online markets for it services. Manage. Sci. **49**(11), 1504–1520 (2003)
22. Zheng, A., Hong, Y., Pavlou, P.: Value uncertainty and buyer contracting: evidence from online labor markets. In: International Conference on Information Systems (ICIS), Fort Worth, Texas, USA (2015)
23. Hong, Y., Pavlou, P.: An empirical investigation on provider pricing in online crowdsourcing markets for IT services. In: Thirty Third International Conference on Information Systems (ICIS), Orlando, USA (2012)
24. Diller, H.: Customer loyalty: fata morgana or realistic goal? Managing relationships with customers. In: Hennig-Thurau, T., Hansen, U. (eds.) Relationship Marketing: Gaining Competitive Advantage Through Customer Satisfaction and Customer Retention, pp. 29–48. Springer, Heidelberg, Germany (2000). https://doi.org/10.1007/978-3-662-09745-8_2
25. Gefen, D.: Customer loyalty in e-commerce. J. Assoc. Inf. Syst. **3**(1), 27–51 (2002)
26. Luarn, P., Lin, H.H.: A customer loyalty model for e-service context. J. Electron. Commer. Res. **4**(4), 156–167 (2003)

27. Zahedi, F.M., Bansal, G., Ische, J.: Success factors in cooperative online marketplaces: trust as the social capital and value generator in vendors—exchange relationships. J. Organ. Comput. Electron. Commer. **20**(4), 295–327 (2010)

28. Malone, T.W., Yates, J., Benjamin, R.I.: Electronic markets and electronic hierarchies. Commun. ACM **30**(6), 484–497 (1987)

29. Kim, J.Y., Wulf, E.: Move to depth: buyer–provider interactions in online service marketplaces. e-Service J. **7**(1), 2–14 (2010)

30. Kim, J.Y.: Online reverse auctions for outsourcing small software projects: determinants of vendor selection. e-Service J. **6**(3), 40–55 (2009)

31. Smith, R.E., Swinyard, W.R.: Information response models: an integrated approach. J. Mark. **46**(1), 81–93 (1982)

32. Jap, S.D.: Online reverse auctions: issues, themes, and prospects for the future. J. Acad. Mark. Sci. **30**(4), 506–525 (2002)

33. Pavlou, P.A., Dimoka, A.: The nature and role of feedback text comments in online marketplaces: implications for trust building, price premiums, and seller differentiation. Inf. Syst. Res. **17**(4), 392–414 (2006)

34. Gefen, D., Carmel, E.: Does reputation really signal potential success in online marketplaces, or is it only a trigger? In: Mediterranean Conference on Information Systems (MCIS), Tel Aviv, Israel (2010)

35. Radkevitch, U., van Heck, E., Koppius, O.: Portfolios of buyer-supplier exchange relationships in an online marketplace for IT services. Decis. Support Syst. **47**(4), 297–306 (2009)

36. Kim, J.Y., Wulf, E.: Service encounters and relationships: buyer–supplier interactions in online service marketplaces. In: Americas Conference on Information Systems (AMCIS), San Francisco, California, USA (2009)

37. Holthaus, C., Stock, R.M.: Good signals, bad signals: performance and trait implications of signaling in online labor markets. In: International Conference on Information Systems (ICIS), Seoul, South Korea (2017)

38. Hu, X., Wu, G., Wu, Y., Zhang, H.: The effects of web assurance seals on consumers' initial trust in an online vendor: a functional perspective. Decis. Support Syst. **48**(2), 407–418 (2010)

39. Clemons, E.K., Wilson, J., Matt, C., Hess, T., Ren, F., Jin, F., Koh, N.S.: Global differences in online shopping behavior: understanding factors leading to trust. J. Manage. Inf. Syst. **33**(4), 1117–1148 (2016)

40. Kim, D.J., Ferrin, D.L., Rao, H.R.: Trust and satisfaction, two stepping stones for successful e-commerce relationships: a longitudinal exploration. Inf. Syst. Res. **20**(2), 237–257 (2009)

41. Gefen, D., Karahanna, E., Straub, D.W.: Trust and TAM in online shopping: an integrated model. MIS Q. **27**(1), 51–90 (2003)

42. Guo, W., Straub, D., Han, X., Zhang, P.: Understanding vendor preference in the crowdsourcing marketplace: the influence of vendor-task fit and swift trust. J. Electron. Commer. Res. **18**(1), 1–17 (2017)

43. Wang, J.-C., Chiang, M.-J.: Social interaction and continuance intention in online auctions: a social capital perspective. Decis. Support Syst. **47**(4), 466–476 (2009)

44. Palvia, P.: The role of trust in e-commerce relational exchange: a unified model. Inf. Manag. **46**(4), 213–220 (2009)

45. Gao, G., Greenwood, B.N., Agarwal, R., McCullough, J.S.: Vocal minority and silent majority: how do online ratings reflect population perceptions of quality? MIS Q. **39**(3), 565–589 (2015)

46. Dellarocas, C.: Reputation mechanism design in online trading environments with pure moral hazard. Inf. Syst. Res. **16**(2), 209–230 (2005)

47. Dellarocas, C.: The digitization of word of mouth: promise and challenges of online feedback mechanisms. Manage. Sci. **49**(10), 1407–1424 (2003)
48. Yen, C.H., Lu, H.P.: Factors influencing online auction repurchase intention. Internet Res. **18**(1), 7–25 (2008)
49. Kim, M.S., Ahn, J.H.: Comparison of trust sources of an online market-maker in the e-marketplace: buyer's and seller's perspectives. J. Comput. Inf. Syst. **47**(1), 84–94 (2006)
50. Chiou, J.S., Wu, L.Y., Sung, Y.P.: Buyer satisfaction and loyalty intention in online auctions: online auction web site versus online auction seller. J. Serv. Manage. **20**(5), 521–543 (2009)
51. Srinivasan, S.S., Anderson, R., Ponnavolu, K.: Customer loyalty in e-commerce: an exploration of its antecedents and consequences. J. Retail. **78**(1), 41–50 (2002)
52. Qi, H., Mao, J.: Facilitating transactions on a crowdsourcing platform: a cognitive frame perspective. In: International Conference on Information Systems (ICIS), Dublin, Ireland (2016)
53. Lu, B., Hirschheim, R.: Online sourcing: investigations from service clients' perspective. In: Americas Conference on Information Systems (AMCIS), Detroit, Michigan, USA (2011)
54. Tomlinson-Keasey, C.: Opportunities and challenges posed by archival data sets. In: Funder, D.C., et al. (eds.) Studying Lives through Time: Personality and Development, pp. 65–92. American Psychological Association, Washington, DC, USA (1993)
55. Elance.com. Elance Online Employment Report, Quarter 3 2011 (2011). https://www.elance.com/q/oer-test. Accessed 19 July 2012
56. Misra, R.B.: Global IT outsourcing: metrics for success of all parties. J. Inf. Technol. Cases Appl. **6**(3), 21–34 (2004)
57. Dibbern, J., Goles, T., Hirschheim, R., Jayatilaka, B.: Information systems outsourcing: a survey and analysis of the literature. Data Base Adv. Inf. Syst. **35**(4), 6–98 (2004)
58. Shen, W.: Essays on Online Reviews: The Relationships between Reviewers, Reviews, and Product Sales, and the Temporal Patterns of Online Reviews (2008)
59. Hennig-Thurau, T., Hansen, U.: Relationship Marketing: Gaining Competitive Advantage through Customer Satisfaction and Customer Retention. Springer, Heidelberg, Germany (2000). https://doi.org/10.1007/978-3-662-09745-8
60. Chin, W.W.: The partial least squares approach to structural equation modeling. In: Marcoulides, G.A. (ed.) Modern Methods for Business Research, pp. 295–336. Lawrence Erlbaum Associates, Mahwah, USA (1998)
61. Hair, J.F., Tomas, G., Hult, M., Ringle, C., Sarstedt, M.: A Primer on Partial Least Squares Structural Equation Modeling (PLS-SEM), 2nd edn. Sage, Thousand Oaks, USA (2016)
62. Urbach, N., Ahlemann, F.: Structural equation modeling in information systems research using partial least squares. J. Inf. Technol. Theory Appl. **11**(2), 5–40 (2010)
63. Haenlein, M., Kaplan, A.M.: A beginner's guide to partial least squares analysis. Underst. Stat. **3**(4), 283–297 (2004)
64. Hair, J.F., Black, W.C., Babin, B.J.: Multivariate Data Analysis: A Global Perspective. Pearson Education, Upper Saddle River, USA (2010)
65. Ringle, C.M., Wende, S., Becker, J.-M.: SmartPLS 3.0. SmartPLS GmbH (2015)
66. Chin, W.W., Marcolin, B.L., Newsted, P.R.: A partial least squares latent variable modeling approach for measuring interaction effects: results from a Monte Carlo simulation study and an electronic-mail emotion/adoption study. Inf. Syst. Res. **14**(2), 189–217 (2003)
67. Henseler, J., Fassott, G.: Testing moderating effects in PLS path models: an illustration of available procedures. In: Esposito Vinzi, V., Chin, W., Henseler, J., Wang, H. (eds.) Handbook of Partial Least Squares, pp. 713–735. Springer, Heidelberg, Germany (2010). https://doi.org/10.1007/978-3-540-32827-8_31

68. Henseler, J., Chin, W.W.: A comparison of approaches for the analysis of interaction effects between latent variables using partial least squares path modeling. Struct. Eqn. Model. Multi. J. **17**(1), 82–109 (2010)

69. Neuman, W.L.: Social Research Methods: Qualitative and Quantitative Approach, 6th edn. Pearson Education, Boston, USA (2006)

70. Limayem, M., Hirt, S.G., Cheung, C.M.K.: How habit limits the predictive power of intention: the case of information systems continuance. MIS Q. **31**(4), 705–737 (2007)

71. Conway, J., Lance, C.: What reviewers should expect from authors regarding common method bias in organizational research. J. Bus. Psychol. **25**(3), 325–334 (2010)

72. Malhotra, N.K., Kim, S.S., Patil, A.: Common method variance in is research: a comparison of alternative approaches and a reanalysis of past research. Manage. Sci. **52**(12), 1865–1883 (2006)

73. Cohen, J.: Statistical Power Analysis for the Behavioral Sciences, 2nd edn. Lawrence Erlbaum Associates, Hillsdale, USA (1988)

74. Lasrado, L.A., Lugmayr, A.: Crowdfunding in Finland: a new alternative disruptive funding instrument for businesses. In: Proceedings of International Conference on Making Sense of Converging Media, pp. 194–201. ACM, Tampere, Finland (2013)

75. Zanatta, A.L., Machado, L.S., Pereira, G.B., Prikladnicki, R., Carmel, E.: Software crowdsourcing platforms. IEEE Softw. **33**(6), 112–116 (2016)

76. Kanat, I.E., Hong, Y., Raghu, T.S.: Surviving in global online labor markets for IT services: a geo-economic analysis. Inf. Syst. Res. **29**, 893 (2018)

77. Ghezzi, A., Gabelloni, D., Martini, A., Natalicchio, A.: Crowdsourcing: a review and suggestions for future research. Int. J. Manage. Rev. **20**(2), 343–363 (2018)

Using NSIA Framework to Evaluate Impact of Sentiment Datasets on Intraday Financial Market Measures: A Case Study

Islam Al Qudah[(⊠)] and Fethi A. Rabhi

School of Computer Science and Engineering,
University of New South Wales, Sydney 2052, Australia
i.al-qudah@unswalumni.com, fethir@cse.unsw.edu.au

Abstract. Current studies on financial markets reaction to news show lack of flexibility for conducting news sentiment datasets evaluations. In other words, there is an absence of clear step-by-step guidance for conducting impact analysis studies in various financial contexts. This paper evaluates the proposed News Sentiment Impact Analysis (NSIA) framework using a highly sensitive financial market measure called the intraday mean cumulative average abnormal returns. The results demonstrate the ability of the framework to evaluate news sentiment impact on high frequency financial data (minutes intervals), while defining clear steps to conduct a systematic evaluation.

1 Introduction

Financial markets reaction to news has been studied extensively, starting few decades ago [1]. Researchers have diverged in their investigation approaches, some focusing on studying the investor sentiment beliefs about companies' cash flows and associated risk in investing in these companies, using the facts they have in hand [2], and how these beliefs impact their decision making. Other researchers focused on studying text sentiment, weighing the subjectivity elements found in text. Between investor and text-based sentiment, this paper is concerned with the latter, as it usually captures investors' subjectivity (such as in tweets and blogs) and the subjectivity found in more formal sources such as news articles and corporate disclosers. The News Sentiment Impact Analysis (NSIA) framework introduced earlier in [3], has proposed a model and case study to evaluate sentiment datasets impact on daily financial market measure. This paper pushes the boundaries little further and tests the framework using more news sensitive market measure like intraday market returns.

The paper is structured as follows. The next section gives some background on sentiment analysis and presents various approaches for measuring the impact on financial markets. Section 3 describes the News Sentiment Impact Analysis Framework, designed for the purpose of carrying out impact analysis evaluation of any given sentiment dataset. Section 4 describes a case study to evaluate the framework. Finally, Sect. 5 concludes this paper.

© Springer Nature Switzerland AG 2019
N. Mehandjiev and B. Saadouni (Eds.): FinanceCom 2018, LNBIP 345, pp. 101–117, 2019.
https://doi.org/10.1007/978-3-030-19037-8_7

2 Background and Related Work

Researchers utilise various sources of information such as corporate disclosers (e.g. [4–6, 15]), tweets (e.g. [7, 8]), Facebook posts, news articles (e.g. [9, 10]), internet message boards and blogs (e.g. [11, 12]). Different sentiment analysis approaches have been employed to extract and determine the subjectivity weight of such text sources. These approaches are dictionary-based approaches (e.g. [6, 9, 10, 13, 14]), machine learning approaches and natural language processing approaches (e.g. [11, 12, 16–20]). In addition, to evaluate the effectiveness of their proposed sentiment analysis models, researchers have utilised various financial markets impact models. Regression analysis methods (e.g. [9–12, 20, 28, 29]) were used in many studies, while trading strategies were trialled by others (e.g. [4, 6, 14, 17]). With the large amount of text involved and the mixture of models been used, conducting such studies by finance researchers has become a data intensive activity which requires a mixture analytic modelling and software analysis skills.

This paper reviewed many related studies in the domain of sentiment driven impact analysis related to finance (see Table 1 Sentiment analysis approaches and impact models). These studies have followed one or a combination of two main methodologies to evaluate the impact of the sentiment data on the market data. These are: regression analysis (e.g. Linear Regression (LnR), Ordinary Least Squares (OLS) regression, Logistic Regression (LoR)) and trading strategies.

Table 1. Sentiment analysis approaches and impact models

Research study		Sentiment analysis approach		Impact model			
		Machine learning	Knowledge based	Regression analysis			Trading strategy
				LnR	LoR	OLS	
1	[11]	✓		✓			
2	[4]	✓					✓
3	[12]	✓		✓			
4	[9, 10]		✓	✓			✓
5	[13]		✓	✓			✓
6	[5]		✓	✓			✓
7	[21]		✓	✓			
8	[22]		✓	✓			
9	[23]		✓	✓			
10	[7]	✓		✓			
11	[16]	✓				✓	
12	[14]		✓	✓	✓		✓
13	[45]		✓	✓			
14	[46]		✓	✓			

(continued)

Table 1. (*continued*)

Research study		Sentiment analysis approach		Impact model			
		Machine learning	Knowledge based	Regression analysis			Trading strategy
				LnR	LoR	OLS	
15	[24]		✓	✓			
16	[25]		✓	✓			✓
17	[26]		✓	✓			
18	[17]	✓					✓
19	[18]	✓		✓			✓
20	[19]	✓		✓			✓
21	[8]	✓			✓		
22	[27]	✓			✓	✓	
23	[28]		✓	✓			
24	[29]		✓	✓			✓
25	[6]		✓				✓
26	[20]	✓				✓	
27	[15]		✓	✓			

The reviewed studies show that researchers have utilised a variety of sentiment analysis techniques and impact models. One group of studies (mainly computing extensive based studies) focused on proposing techniques which encapsulate processes to perform sentiment analysis for a specific text source, with limited impact analysis for validation purposes. The researchers in this group have advanced computing skills. Their research activities mainly focus around developing new sentiment analysis algorithms (i.e. machine learning). However, the corresponding impact measuring processes are often not documented and only accessible by the authors of the study. Users who wish to reuse part or all of their evaluation processes in different financial contexts, or for different news sources, hit a roadblock. While studies with strong finance background and domain expertise in financial markets modelling and evaluation, have focused their work on evaluating the impact of a particular sentiment dataset using different financial market measures. Many of the evaluation activities are hard to reproduce for different sentiment datasets, as many of these activities are not automated. In addition, evaluation requires users to dedicate a great portion of their time to conduct these activities.

In conclusion, most of the results described in the literature concerning sentiment analysis perform impact analysis in a way that is difficult to reproduce outside of a specific context, i.e. the processes involved in evaluating the impact of a sentiment dataset are not documented. To reproduce a studies' results, one would need to implement the evaluation algorithms and have access to the input (news and market data) used. This makes the task of reproducing results a complex job ([30]; Peng 2011). The impact on financial markets can be gauged by several measures, but the majority of the studies reviewed focused on studying the impact using one or two financial market measures. Stock returns (intraday, daily, monthly, annually) have been studied widely, and considered to be the most popular performance measure of impact on financial markets. However, the literature shows there are many other measures that could be indicators for impact such as liquidity, volatility [31]. The complexity of sentiment analysis models along with time limitation and/or limited knowledge in the financial markets domain, leads researchers to apply limited impact analysis evaluation.

The existing studies use a variety of statistical methods to test the impact sentiment had on financial market measures. Some use correlation tests, while others use regression analysis to discover the strength of the relationship between change in market measures and sentiment datasets. Others take one step further, implementing several regressions between the different financial market measures. In addition, most studies evaluated their impact models against one specific fixed financial context. However, this raises a question like *"what if the financial context or the financial market measure changes? would I get the same results?"*. Automating the processes involved in analysing and evaluating data, whether it is market data, or sentiment data has many advantages: first reducing labour/manual work, reducing human errors, and reducing time and resources [32].

3   News Sentiment Impact Analysis (NSIA) Framework

The research gap has three different aspects: first aspect relates to the shortcomings of existing literature, where the lack of *flexibility* for conducting news sentiment datasets evaluations has been noticed. The second aspect relates to the absence of clear *step-by-step* guidance for replicating the results, i.e. little guidance on reproducing or reconducting a set of experiments. Therefore, there is no systematic methodology that has clear step-by-step use cases to enable *reproducibility* and *consistency* in conducting sentiment-driven impact analysis studies. The third aspect relates to lack of software tools to support *automation* of such step-by-step use cases. The role of such tools would be to allow experiments to be conducted effectively and minimize the risk of errors.

To address this research problem the paper utilises a software framework called News Sentiment Impact Analysis (NSIA) which is comprised of three components: the first component is a data model called the NSIA data model which consists of Market Data (MD) model, Sentiment Data (SD) model and a novel data model called the Comparison Parameters Data (CPD) model that captures contextual parameters in the financial markets and news sentiment analysis sphere. The second component proposes a set of step-by-step predefined *use cases* to make the job of conducting experiments repeatable for the users. This will address the *reproducibility* issue by allowing users to repeat impact analysis studies. The third component is a *software architecture* that is designed to support both the data model, and the use cases associated with it. The software architecture should be able to facilitate *automation* of the impact analysis studies. The architecture also facilitates the reuse and *interoperability* of an existing software components, libraries, and packages in conducting impact analysis studies. The rest of this section describes these components in more detail.

3.1 NSIA Data Model

The NSIA data model is made up of three distinct models to incorporate Market Data (MD), Sentiment Data (SD) and Comparison Parameters Data (CPD).

3.1.1 Market Data (MD) Model

The Market Data model represents the datasets provided by financial market data providers. This model is generic and flexible enough to capture any dataset originating from providers such as Thomson Reuters [33] and Bloomberg [34]. The conceptual data model is adapted from [35, 36] and is composed of number of *entities*, which are described as follows:

- Event: time-stamped superclass capturing different types of events. It could be extended to represent any event across different domains, for example news events.
- Product: Products are distinguished by *ProductID* key. There are two types of products: tradable and non-tradable products. Trades, quotes, end of day and market depth are tradable events, their *ProductID* would be the code of the company issuing these products. Non-tradable products include index, news and measure events. An index event's *ProductID* would be the index code.
- *Exchange*: provide platforms to trade products. Companies list and trade their products on the exchange. Exchanges maintain market datasets in either high-frequency form [37], or in low-frequency form as in end of day transactions.
- The market data model defines End of Day, Quote, Trade, Market Depth, Index and Market Measure events. These are described as follows:
- *End of Day*: timestamped events that represent values of trades on daily basis. Those include: Opening Price, Closing Price, Highest and Lowest Price values for a particular *ProductID*.
- *Quote*: timestamped event that lists the best bid and ask submitted to the exchange by market participants (brokers, traders) for a particular *ProductID*.
- *Trade*: timestamped events, which show the trades that took place, for a particular *ProductID*.
- *Index*: timestamped events that represent the value of a particular index.
- *Market Depth*: timestamped events showing the depth and breadth of quotes events up to a certain level (e.g. 10th best bid and 10 best ask for a particular *ProductID*).
- *Market Measure*: market measure events store timestamped data related to different measures for a particular *ProductID* such as: *Liquidity, Volatility, Intraday Returns* and *Daily Returns*.

3.1.2 Sentiment Data (SD) Model

The sentiment data model is designed to represent news and sentiment datasets. The proposed model extends the event superclass with two events *News Items* and *News Analytics*. The following is a description of these entities:

- News Item: timestamped event that represents a news story, which is issued on a scheduled or unscheduled basis. News items store information such as news headline, news keywords, news topics, news body and news release date and time.

- News Analytics: timestamped news sentiment record. This is another type of event that carries additional information about the news record. For instance, it stores sentiment related information, the news novelty and news relevance scores.

3.1.3   Comparison Parameters Data (CPD) Model

The CPD model is the key component of the NSIA data model as it represents contextual parameters associated with conducting sentiment-driven impact analysis studies.

3.1.4   Defining the CPD Model

The Comparison Parameters Data (CPD) model divides the contextual parameters into three sets: Financial market context parameters (FC), Sentiment extraction parameters (SN) and Impact Measure parameters (IM) as described in details in [3].

Financial Context (FC) Parameters

The *FC_PARAM* entity represents a given financial context and is linked to entities in the market data model through the following relationships:

- CtxEntity: This relationship links financial entities (such as company ids) that are part of the impact study to the corresponding Product entity (Fig. 1).

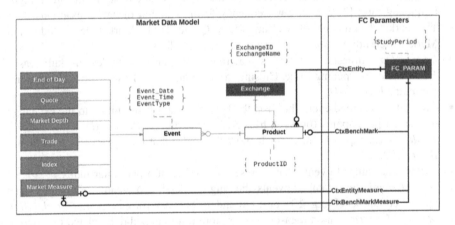

Fig. 1. Defining financial context model

- CtxEntityMeasure: This relationship enables associating the financial context with a MarketMeasure event. A measurable event could be intraday return, daily returns … etc.
- CtxBenchmark: This relationship defines a benchmark that is offered as a product in the Product entity, usually an Index event. Defining benchmark entities is a method used in many event data evaluation studies [44].
- CtxBenchmarkMeasure: This relationship associates a benchmark with a market measure event in the MarketMeasure entity. For instance, it could represent an aggregate value of an event, say for example a business sector value, an interest rate value, or an index value.

The *FC_PARAM* entity has the following attributes:

- StudyPeriod: This attribute represents the date or time ranges of events. It is mapped to EventDate and EventTime attribute in Event entity.

Sentiment Extraction parameters (SN)
The SN parameters in the CPD model define five entities as shown in Fig. 2, to filter news and identify sentiment datasets relevant for the study. These are explained as follows:

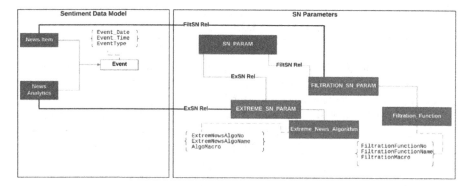

Fig. 2. Defining SN parameter entities

- SN_PARAM: the main entity which acts as a root for the other entities in the SN parameters model.
- FILTRATION_SN_PARAM and Filtration_Function: the first entity links to the News Item instances via the relationship FiltSN Rel. These instances are produced by the function defined in the Filtration_Function entity.
- EXTREME_SN_PARAM and Extreme_News_Algorithm: the first entity links to the News Analytics instances via the relationship ExtSN Rel. These instances are produced by the algorithm defined in the Extreme_News_Algorithm entity.

Impact Measure Parameters (IM)
The CPD model defines *IM_PARAM* and *IMPACT_MEASURE* entities, which enable the CPD model to apply different impact models (see Fig. 3).
   These entities are defined as follows:

- IM_PARAM: the root entity which connects the StudyNo with the impact measure used in that study.
- IMPACT_MEASURE: this is a superclass which facilitates the implementation of various impact models, as discussed in Subsect. 3.1.4.

The CPD model provides four impact measures that illustrate the variety of impact models that could be represented in the CPD model. These four impacts measures are:

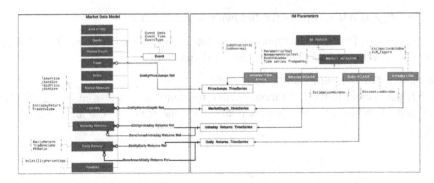

Fig. 3. Defining impact measures parameters

- Daily Mean Cumulative Average Abnormal Returns (Daily MCAAR): this impact measure uses an event study methodology explained in [48], to calculate the daily mean cumulative average abnormal returns.
- Intraday Mean Cumulative Average Abnormal Returns (Intraday MCAAR): this measure is applied to intraday data. It uses high frequency returns time series to calculate the intraday (e.g. 5-min, 10-min intervals) mean cumulative average abnormal returns methodology as demonstrated in [18].
- Intraday Price Jumps: measures the volatilities in stock price timeseries data. The measure used based on the method proposed in Lee and Mykland [38] and used in Bohn et al. [44], which is capable of capturing the timing and size of price jumps. The measure applies a threshold over the stock prices observations and a price jump is recorded if an observation breaks through the threshold.
- Intraday Liquidity Based Model (Intraday LBM): this measure uses the EXchange Liquidity Measure (XLM) method, which calculates the trading costs of a roundtrip trade of a given size as explained in Gomber et al. [39].

Accordingly, a number of time series entities which preserve timeseries data calculated using these impact models, are defined as subclasses of the *IMPACT_MEA-SURE* entity. These entities are:

- Daily_Returns_TimeSeries: This entity preserves instances of Daily Returns dataset via EntityDaily Returns Rel and BenchmarkDaily Returns Rel relationships.
- Intraday_Returns_TimeSeries: This entity preserves instances of Intraday Returns dataset via EntityIntraday Returns Rel and BenchmarkIntraday Returns Rel relationships.
- PriceJumps_TimeSeries: This entity preserves instances of Trade dataset via EntityPriceJumps Rel relationship.
- MarketDepth_TimeSeries: This entity preserves instances of Liquidity dataset via EntityMarketDepth Rel relationship.

CPD Model Results
The CPD model logs the set of parameters selected for each impact study using the *CPD_PARAM_LOG* entity, distinguishing studies by *StudyNo* attribute.

3.2   NSIA Architecture

The NSIA architecture is designed to support the proposed data model and the impact analysis use cases. The NSIA architecture follows the ADAGE Framework guidelines [40] and is designed using a combination of both component based and service-oriented design principles. The architecture encompasses three layers: GUI layer, Business layer and Data layer. These layers are summarized as follows:

- GUI layer: mediates the interactions between users and the Business layer, based on user selections. The user interfaces provided in the NSIA architecture enable actionizing the data model defined in Sect. 3.1. Users through the GUI layer invoke use cases defined in Sect. 3.3.
- Business layer consolidates a number of components which encapsulate the majority of the framework's business logic.
- Data layer: a number of data repositories are used to cater for the complete cycle of conducting sentiment-driven impact analysis studies.

3.3   NSIA Use Cases

The third design artefact introduces a number of *use cases* to guide the user in conducting sentiment-driven impact analysis studies using the NSIA data model and architecture.

3.3.1   Overview of Use Cases

The NSIA framework define three use cases, which are:

- Define financial context parameters: this use case guides the analyst/user in defining the financial context parameters.
- Define Sentiment extraction parameters: this use case assists the analyst/user in defining the sentiment extraction parameters.
- Conduct impact analysis: this use case assists the user in defining the Impact Measure parameters and conducting impact analysis studies.

The sequence diagrams that correspond to these three use cases are now described in more detail in the rest of the section.

3.3.2   Define Financial Context Parameters

This use case enables users to set the FC parameters via a GUI and load the market data needed for the evaluation. The Data Model Management (DMM) component is responsible for logging the FC parameters in the CPD database, generating a study number for the new impact study and for creating market data subsets relevant to the FC parameters, then log them to the CPD database.

3.3.3   Define Sentiment Extraction Parameters

This use case enables users to set the SN parameters via a GUI and load the sentiment data needed for the evaluation. The Data Model Management (DMM) component is responsible for logging the SN parameters in the CPD database and invoking the Sentiment Processing (SP) component, which filters news sentiment (according to

Filtration function FA) and creates subsets of extreme news sentiment datasets (according to the Extreme Sentiment Extraction (ESE) algorithm) in the CPD database.

3.3.4   Conduct Impact Analysis

The impact analysis use case assumes that FC and SN parameters have been defined and that market and sentiment datasets, which are the subject of the evaluation have been identified. The use case sequence diagram is shown in Fig. 4. It starts when the user defines the appropriate Impact Measure (IM) parameters, which are saved in the CPD database by the DMM component. The user then invokes Impact Analysis (IA) component via a GUI, which retrieves the relevant market and sentiment subsets as per step 6 in Fig. 4. The IA component applies statistical significance tests to compute the impact results as per step 8. The IA component then logs the impact results to the CPD database.

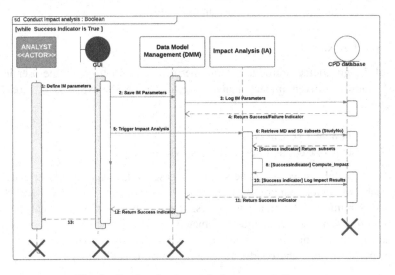

Fig. 4. Conduct impact analysis sequence diagram

4   Case Study: Evaluating Impact of Negative News Sentiment Dataset Using Intraday Market Measures

The NSIA framework has been evaluated previously using two case studies, which were published in in [3]. These studies tested the selected commercial sentiment dataset's impact on daily abnormal returns. This case study evaluates the same sentiment dataset using intraday abnormal returns, which is more news sensitive market measure. We first define the CPD model parameters, then perform the use cases, and finally discuss the results. The choice of negative news dataset over positive news, has been proved by researchers to show that negative news do in fact affect markets more than positive news [47].

4.1   Defining the CPD Model Parameters

The financial contexts parameters and sentiment parameters selected for this case study are similar to the second case study introduced in [3]. Next, the case study uses three *intraday* Mean Cumulative Average Abnormal Returns (MCAAR) as a measure of impact. The parameters used by this impact measure are shown Table 2.

Table 2. Defining the IM parameters

Impact measure parameter	Impact measure Type	Parameter name	Parameter value	Description
IM parameter	Intraday MCAAR	Estimation window	(−17 Days, −3 Days)	Denotes the period of 14 days before the news release date. This window is used to calculate the expected returns as clarified in the implementation of this model
Event window			{(−40 min, −10 min), (0 min, 30 min)}	0 denotes the time of news release
Time series frequency			Five-minute	Determine the frequency of the generated timeseries data

The case study executed the three use cases defined briefly in Sect. 3. These use cases define the FC and the SN parameters and then conduct impact analysis.

4.2   Results Discussion

The results that correspond to selecting intraday MCAAR as the impact model are shown in Table 3. Column Country stores the country of the market index and its constituents. Column Benchmark show the abbreviation for the market index name. Column |D| is the number of distinct event days decided by applying the ESE algorithm (ESE_VOL and ESE_TOT). The number of extreme news can be different from the number of distinct days, as it is possible to have multiple extreme news in the same day. Columns MCAAR (−40 min, −10 min) and MCAAR (0 min, 30 min) are two event windows that show the mean intraday MCAAR during these two windows. The statistical significance is tested using parametric Welch two sample t-test, which gives two figures (columns Welch two sample t-test and P value). The statistical significance of the results is highlighted with symbols \$, *, **, and ***, which denote the statistical significance at the 0.10, 0.05, 0.01 and 0.001 levels, respectively.

In nine out of the twelve studies, intraday MCAAR dropped after releasing the news (window 0 to 30 min) compared with before releasing the news (window −40 to −10 min). In eight studies (study 3, 4, 7, 8, 9 and 11), the drop was significant at various levels. This provides enough evidence that there is strong correlation between the news filtered by ESE_VOL and ESE_TOT algorithms and the intraday (5 min interval) impact witnessed after the time of releasing the news. Figure 5 provides a sample visualization of impact magnitude over a period of 1 h 30 min before and 30 min after the new release time for 5 min intervals.

Negative news volumes (ESE_VOL) algorithm generated better impact results than negative news weights (ESE_TOT) algorithm. This is evident in all contexts studied (across all countries). In addition, the number of days with significant high volumes is smaller than the number of days with lower volumes and higher sentiment weights (sentiment scores). This observation is true, except in the case of the USA markets, where the number of days with high volumes was higher than the number of days with extreme negative sentiment scores (ESE_TOT).

Table 3. Intraday MCAAR impact results

Study no	Country	Benchmark	ESE Algo.	|ESP|	|D|	MCAAR (−40 min, −10 min) %	MCAAR (0 min, 30 min) %	Welch Two Sample T-Test	P Value
1	Australia	ATLI	ESE_VOL	30	5	3.65	4.47	−1.49	0.9067
2	Australia	ATLI	ESE_TOT	28	12	1.12	2.18	−3.74	0.9979
3	Australia	AORD	ESE_VOL	30	5	6.13	−6.97	4.51	0.00201**
4	Australia	AORD	ESE_TOT	28	12	1.23	0.38	6.79	0.00006***
5	Germany	GDAXI	ESE_VOL	37	6	−25.23	−24.79	−4.14	0.9992
6	Germany	GDAXI	ESE_TOT	34	9	3.66	3.7	−0.83	0.7857
7	Canada	SPTSE	ESE_VOL	2	1	2.99	2.47	3.04	0.01431**
8	Canada	SPTSE	ESE_TOT	10	7	2.33	1.54	8.89	0.00002***
9	USA	DJI	ESE_VOL	28	10	4.53	3.29	6.39	0.00008***
10	USA	DJI	ESE_TOT	15	6	9.63	9.45	2.13	0.06031$
11	USA	HWI	ESE_VOL	28	10	7.21	5.86	4.99	0.00101***
12	USA	HWI	ESE_TOT	15	6	8.87	8.63	1.77	0.07433$

The results confirm that the ESE_VOL algorithm shows better results than the ESE_TOT algorithm, where out of the six impact studies using ESE_VOL, four generated statistically significant results (column P-Value). This result complies with findings in literature, that news volume is correlated with higher/lower significant returns (Das and Chen 2007). Impact studies using ESE_TOT algorithm generated significant results in also four out of the six studies (studies 4, 8, 10 and 12), with less significance. Out of four countries included in the case study, only Germany related studies didn't reflect any level of significance in their results (studies 5 and 6). This could be explained that german investors tend to take a long-term view about the future prospects of firms and do not react to short-term changes in investors' sentiment. This finding shows the importance of validating sentiment datasets across different financial contexts (countries, benchmarks), where assumptions made on the efficiency/deficiency of a sentiment dataset, could be confirmed/refuted, when testing against other financial contexts.

4.3   Discussion

This case study focused on improving impact analysis by evaluating an intraday impact model (on 5-min interval timeseries. As expected, the impact results were much more significant than those of the previous case studies. They show the immediate impact of releasing the news with a short time period (30 min after releasing the negative news), where significant drops in returns were observed.

In addition, the results enabled the comparison between daily and intraday impact magnitude. Table 4 compares the *daily* MCAAR (presented in [3]) and *intraday* MCAAR impact figures presented in this case study. Column Intraday MCAAR Impact Magnitude show the difference between MCCAR (0 min, 30 min) figures and MCAAR (−40 min, −10 min) figures, generated in the third case study. Column P-value represents the statistical significance of the figures shown in the Intraday MCAAR Impact

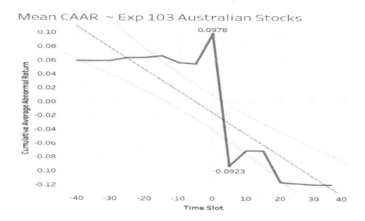

Fig. 5. Intraday MCAAR results for study number 3

Table 4. Intraday vs Daily MCAAR results

Study no	Country	Benchmark	ESE Algo.	Intraday Impact Results					Daily Impact Results	
				MCAAR (−40 min, −10 min) %	MCAAR (0 min, 30 min) %	Intraday MCAAR Impact Magnitude %	Welch Two Sample T-Test	P Value	Daily MCAAR Impact Magnitude %	Generalized Sign Z
1	Australia	ATLI	ESE_VOL	3.65	4.47	0.82	−1.49	0.9067	−0.03	−0.216
2	Australia	ATLI	ESE_TOT	1.12	2.18	1.06	−3.74	0.9979	−1.05	−1.870*
3	Australia	AORD	ESE_VOL	6.13	−6.97	−13.10	4.51	0.00201**	−0.25	0.265
4	Australia	AORD	ESE_TOT	1.23	0.38	−0.85	6.79	0.00006***	−0.82	−1.383$
5	Germany	GDAXI	ESE_VOL	−25.23	−24.79	0.44	−4.14	0.9992	−0.31	0.926
6	Germany	GDAXI	ESE_TOT	3.66	3.7	0.04	−0.83	0.7857	0.28	0.622
7	Canada	SPTSE	ESE_VOL	2.99	2.47	−0.52	3.04	0.01431**	−1.23	−0.874
8	Canada	SPTSE	ESE_TOT	2.33	1.54	−0.79	8.89	0.00002***	−0.07	1.353$
9	USA	DJI	ESE_VOL	4.53	3.29	−1.24	6.39	0.00008***	−0.10	0.575
10	USA	DJI	ESE_TOT	9.63	9.45	−0.18	2.13	0.06031$	0.69	−0.265
11	USA	HWI	ESE_VOL	7.21	5.86	−1.35	4.99	0.00101***	0.32	2.953**
12	USA	HWI	ESE_TOT	8.87	8.63	−0.24	1.77	0.07433$	0.95	1.482$

Magnitude column, Column Daily MCAAR Impact Magnitude show the daily figures generated in the second case study. Like the P-value column, column Generalized Sign Z determines the statistical significance of the daily impact figures. The statistical significance is highlighted with symbols $, *, **, and ***, which denote the statistical significance at the 0.10, 0.05, 0.01 and 0.001 levels, respectively.

In the light of the results, several observations can be summarised as follows:

- The table confirms that daily MCAAR figures show weaker signs of reaction to the negative news sets identified as extreme.
- There are cases in which daily figures don't reveal any impact at all. For example, in studies 9 and 10 related to Dow Jones Index, the impact of the daily MCAAR figures showed no signs of reaction to news, while the intraday impact magnitude and the p-values showed more reliable immediate impact of news on the abnormal returns. The daily MCAAR impact magnitude and Generalized Sign Z columns show the markets absorbed the negative news and by the end of the day the abnormal returns evaporated as compared to time periods around the news time.
- There are cases where the intraday MCAAR figures showed no signs of reaction to negative sentiment in the news, while the daily impact figures showed negative MCAAR figures e.g. studies 2 and 5 related to Australia's ASX top 20 leaders and Germany's DAX Index respectively. We can't attribute the impact in these cases to the negative news, as the intraday figures showed no signs of impact, this could mean the daily figures are just by chance, or there could be other factors contributing to these results. In conclusion, relying on the daily impact results in these cases could be misleading and not a true reflection of the real impact.

5   Conclusion and Future Work

This paper evaluated the News Sentiment Impact Analysis (NSIA) Framework using intraday market measures. The case study results reflect how changing the context parameters impact the MCAR results. Although the results are interesting, the NSIA framework need to be tested more robustly, via Introducing new use cases to increase automation and allow large scale impact studies to be conducted. Such use cases can provide end to end analytics which would include importing market and sentiment data on the fly, evaluating thousands of financial contexts against one sentiment dataset, and allow the user to interact visually with the results. Additionally, future work involves incorporating additional sentiment datasets into the framework. Like for example the sentiment datasets offered by Quandl [41] and Ravenpack [42]. It would be interesting to provide the users with the ability to compare impact results of more than one sentiment data source, which could bring better understanding how different financial contexts respond to these datasets.

Acknowledgments. We are grateful to Sirca [43] and Thomson Reuters [33] for providing access to the data used in this research.

References

1. Niederhoffer, V.: The analysis of world events and stock prices. J. Bus. **44**(2), 193–219 (1971)
2. Baker, M., Wurgler, J.: Investor sentiment and the cross-section of stock returns. J. Finan. **61**(4), 1645–1680 (2006)
3. Qudah, I., Rabhi, F.A.: News sentiment impact analysis (NSIA) framework. In: International Workshop on Enterprise Applications and Services in the Finance Industry, pp. 1–16 (2016)
4. Mittermayer, M.A.: Forecasting intraday stock price trends with text mining techniques. In: Proceedings of the 37th Annual Hawaii International Conference on System Sciences 2004, pp. 10-pp. IEEE, January 2004
5. Feldman, R., Govindaraj, S., Livnat, J., Segal, B.: The incremental information content of tone change in management discussion and analysis (2008)
6. Feuerriegel, S., Neumann, D.: Evaluation of news-based trading strategies. In: International Workshop on Enterprise Applications and Services in the Finance Industry, pp. 13–28 (2014)
7. Bollen, J., Mao, H.: Twitter mood as a stock market predictor. Computer **44**(10), 91–94 (2011)
8. Vu, T.T., Chang, S., Ha, Q.T., Collier, N.: An experiment in integrating sentiment features for tech stock prediction in twitter (2012)
9. Tetlock, P.C.: Giving content to investor sentiment: the role of media in the stock market. J. Finan. **62**(3), 1139–1168 (2007)
10. Tetlock, P.C., Saar-Tsechansky, M., Macskassy, S.: More than words: quantifying language to measure firms' fundamentals. J. Finan. **63**(3), 1437–1467 (2008)
11. Antweiler, W., Frank, M.Z.: Is all that talk just noise? The information content of internet stock message boards. J. Finan. **59**(3), 1259–1294 (2004)
12. Das, S.R., Chen, M.Y.: Yahoo! for Amazon: Sentiment extraction from small talk on the web. Manage. Sci. **53**(9), 1375–1388 (2007)
13. Engelberg, J.: Costly information processing: evidence from earnings announcements (2008)
14. Loughran, T., McDonald, B.: When is a liability not a liability? Textual analysis, dictionaries, and 10-Ks. J. Finan. **66**(1), 35–65 (2011)
15. Davis, A.K., Ge, W., Matsumoto, D., Zhang, J.L.: The effect of manager-specific optimism on the tone of earnings conference calls. Rev. Acc. Stud. **20**(2), 639–673 (2015)
16. Dzielinski, M.: News sensitivity and the cross-section of stock returns. Available at SSRN (2011)
17. Schumaker, R.P., Zhang, Y., Huang, C.N., Chen, H.: Evaluating sentiment in financial news articles. Decis. Support Syst. **53**(3), 458–464 (2012)
18. Siering, M.: "Boom" or" Ruin"–does it make a difference? Using text mining and sentiment analysis to support intraday investment decisions. In: 2012 45th Hawaii International Conference on System Science (HICSS), pp. 1050–1059. IEEE (2012)
19. Siering, M.: Investigating the impact of media sentiment and investor attention on financial markets. In: Rabhi, F.A., Gomber, P. (eds.) FinanceCom 2012. LNBIP, vol. 135, pp. 3–19. Springer, Heidelberg (2013). https://doi.org/10.1007/978-3-642-36219-4_1
20. Allen, D.E., McAleer, M., Singh, A.K.: Daily Market News Sentiment and Stock Prices (No. 15-090/III). Tinbergen Institute Discussion Paper (2015)
21. Henry, E.: Are investors influenced by how earnings press releases are written? J. Bus. Commun. (1973) **45**(4), 363–407 (2008)
22. Henry, E., Leone, A.J.: Measuring qualitative information in capital markets research (2009)

23. Kothari, S.P., Li, X., Short, J.E.: The effect of disclosures by management, analysts, and business press on cost of capital, return volatility, and analyst forecasts: a study using content analysis. Account. Rev. **84**(5), 1639–1670 (2009)

24. Doran, J.S., Peterson, D.R., Price, S.M.: Earnings conference call content and stock price: the case of REITs. J. Real Estate Finan. Econ. **45**(2), 402–434 (2012)

25. Engelberg, J.E., Reed, A.V., Ringgenberg, M.C.: How are shorts informed?: short sellers, news, and information processing. J. Financ. Econ. **105**(2), 260–278 (2012)

26. Price, S.M., Doran, J.S., Peterson, D.R., Bliss, B.A.: Earnings conference calls and stock returns: the incremental informativeness of textual tone. J. Bank. Finance **36**(4), 992–1011 (2012)

27. Hagenau, M., Liebmann, M., Neumann, D.: Automated news reading: stock price prediction based on financial news using context-capturing features. Decis. Support Syst. **55**(3), 685–697 (2013)

28. Jegadeesh, N., Wu, D.: Word power: a new approach for content analysis. J. Financ. Econ. **110**(3), 712–729 (2013)

29. Demers, E.A., Vega, C.: Understanding the role of managerial optimism and uncertainty in the price formation process: evidence from the textual content of earnings announcements (2014)

30. Jasny, B.R., Chin, G., Chong, L., Vignieri, S.: Data replication & reproducibility. Science (New York, N.Y.) **334**(6060), 1225 (2011)

31. Lugmayr, A.: Predicting the future of investor sentiment with social media in stock exchange investments: a basic framework for the DAX performance index. In: Friedrichsen, M., Mühl-Benninghaus, W. (eds.) Handbook of Social Media Management, pp. 565–589. Springer, Heidelberg (2013)

32. Harcar, D.M.: Justification and expected benefits of data analysis automation projects. Retrieved August, 2016. https://www.statsoft.com/Portals/0/Support/Download/White-Papers/Automation-Projects.pdf

33. Thomson Reuters: Thomson Reuters News Analytics(TRNA) (2014). http://thomsonreuters.com/products/financial-risk/01_255/news-analytics-product-brochure–oct-2010.pdf. Accessed Jan 2014

34. Bloomberg: Bloomberg news and stocks data feed (2016). http://www.bloomberg.com/markets/stocks. Accessed Apr 2016

35. Rabhi, F.A., Guabtni, A., Yao, L.: A data model for processing financial market and news data. Int. J. Electron. Finan. **3**(4), 387–403 (2009)

36. Milosevic, Z., Chen, W., Berry, A., Rabhi, F.A.: An open architecture for event-based analytics. Int. J. Data Sci. Anal. **2**(1–2), 13–27 (2016)

37. Tsay, R.S.: Analysis of Financial Time Series, vol. 543. Wiley, Hoboken (2005)

38. Lee, S.S., Mykland, P.A.: Jumps in financial markets: a new nonparametric test and jump dynamics. Rev. Finan. Stud. **21**(6), 2535–2563 (2007)

39. Gomber, P., Schweickert, U., Theissen, E.: Liquidity dynamics in an electronic open limit order book: An event study approach. Eur. Finan. Manag. **21**(1), 52–78 (2015)

40. Rabhi, F.A., Yao, L., Guabtni, A.: ADAGE: a framework for supporting user-driven ad-hoc data analysis processes. Computing **94**(6), 489–519 (2012)

41. Quandl: Quandl AAII investor sentiment data (2016). https://www.quandl.com/data/AAII/AAII_SENTIMENT-AAII-Investor-Sentiment-Data. Accessed Apr 2016

42. RavenPack. (2016) RavenPack. http://www.ravenpack.com/. Accessed Apr 2016

43. Sirca: Thomson Reuters Tick History portal (2017). https://tickhistory.thomsonreuters.com/TickHistory/login.jsp. Accessed June 2017

44. Bohn, N., Rabhi, F.A., Kundisch, D., Yao, L., Mutter, T.: Towards automated event studies using high frequency news and trading data. In: Rabhi, F.A., Gomber, P. (eds.) FinanceCom 2012. LNBIP, vol. 135, pp. 20–41. Springer, Heidelberg (2013). https://doi.org/10.1007/978-3-642-36219-4_2

45. Davis, A.K., Piger, J.M., Sedor, L.M.: Beyond the numbers: measuring the information content of earnings press release language. Contemp. Account. Res. **29**(3), 845–868 (2012)

46. Davis, A.K., Tama-Sweet, I.: Managers' use of language across alternative disclosure outlets: earnings press releases versus MD&A. Contemp. Account. Res. **29**(3), 804–837 (2012)

47. Yu, J., Zhou, H.: The asymmetric impacts of good and bad news on opinion divergence: Evidence from revisions to the S&P 500 index. J. Account. Finan. **13**(1), 89–107 (2013)

48. Agrawal, M., Kishore, R., Rao, H. R.: Market reactions to e-business outsourcing announcements: an event study. Info. Manag. **43**(7), 861–873 (2006)

Financial Data Visualization in 3D on Immersive Virtual Reality Displays

A Case-Study for Data Storytelling and Information Visualization of Financial Data of Australia's Energy Sector

Artur Lugmayr[1(✉)], Yi Juin Lim[1], Joshua Hollick[3], Joyce Khuu[2], and Felix Chan[2]

[1] Visualisation and Interactive Media Lab. (VisLab),
Curtin University, Perth, Australia
artur.lugmayr@artur-lugmayr.com,
lartur@acm.org
[2] Curtin Business School, Curtin University, Perth, Australia
{joyce.khuu, f.chan}@curtin.edu.au
[3] Curtin HIVE (Hub for Immersive Visualisation and eResearch),
Curtin University, Perth, Australia
Joshua.Hollick@curtin.edu.au,
http://curtin.edu/vismedia,
http://www.artur-lugmayr.com,
http://www.curtin.edu.au/HIVE

Abstract. Within the scope of this publication, we present a design prototype for the 3D visualization of financial data of Australia's Energy sector with a large scale immersive virtual reality environment. We review existing 3D visualization approaches in finance industry; discuss the technical setup of the prototype; explore design considerations; and approaches to make financial information understandable for everyone. We conclude with a discussion about the potential that immersive virtual environments provide to support the understanding of market events, as well as the exploration of large multidimensional data-sets.

1 Introduction

Financial market data is complex and requires data presentations that go far beyond simple charts of price data for an instrument. Traditionally financial market data is expressed through dashboards containing charts, news-tickers, trading desks, and other relevant information to support users in gaining an understanding of happenings on the market. Typical visualizations include pricing data, charts, fundamentals, market research, news, alerts, social media data, 3rd party informational content, and enable trading functionality.

Despite the trend towards fully computerized algorithmic trading systems (e.g. [1]) - humans are still a major factor in making trading decisions and are required to quickly

N. Mehandjiev and B. Saadouni (Eds.): FinanceCom 2018, LNBIP 345, pp. 118–130, 2019.
https://doi.org/10.1007/978-3-030-19037-8_8

develop an overview of the current market situations and events taking place. Today's systems present information predominately in the 2D space. We would like to explore the additional opportunities enabled by adding a 3rd dimension for information presentation. We also would like to support humans in gaining easy understanding and knowledge of the current market situation. This also implies faster and better decision making concerning trading decisions. In addition, this will help allow individuals from outside the financial industry to quickly process and understand complex information in a visual manner.

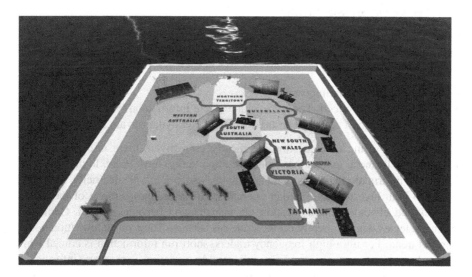

Fig. 1. Screenshot of the final implementation.

To illustrate the potential of 3D, immersive environments, and data storytelling, we have been developing a proof-of-concept implementation we call "ElectrAus", a tool to visualize publicly available data from the Australian energy market. The objective of the proof-of-concept implementation was to illustrate new concepts in data visualization such as:

- Utilization of immersive environments and large screens for financial data visualization;
- Experimentation with data storytelling;
- Creating an understandable visualization for the general public;
- Additional insights and analytics through 3D display;
- Interactivity to create more engagement and allowing to dig in data materials;

1.1   Related Works

Research in financial markets is broad and includes asset pricing, market behavior research, trading algorithms, text/sentiment analysis (see e.g. [2, 3, 4]) and the investigation of social media in trading [5]. Within the scope of this paper, we attempt to apply solutions emerging from utilizing immersive 3D technologies for financial visualization. We give an overview of current research works independently targeted towards investors, operators, or end-users. However, the paper rounds up by presenting our prototype (see Fig. 1) of our prototype, targeted towards end-users. This prototype illustrates the potential of applying 3D visualization for data storytelling in immersive virtual reality environments. It can be considered an example of *Serious Storytelling* – using an environment to tell "stories beyond entertainment purpose" and in serious application contexts [6].

1.2 Theoretical Considerations

Besides the development of a design concept, the main objective of data visualization within financial industries is to enhance effective decision making, based on the following three cornerstones:

- Which information is important?
- What is the most effective way to communicate information?
- How is information accuracy ensured when market dynamics are changing rapidly?

While answers to the first question are mostly user-driven, there is common information in the financial market that would be important to most users. In the context of this paper, information such as demand, supply and pricing of electricity are fundamental to both investors and market operators. However, the structure of the data is complex and effective communication via visualization becomes crucial. Communication strategies depend greatly on the trade-off between short run information versus long run trend. For ultra-high frequency traders, short run information is crucial while for market operators, the intra-daily and weekly variations are more informative. Therefore, an ideal framework should be able to provide "multiple" stories, which allows users to select the "story" that is most *appealing* to them.

2 Method and Approach

For the purpose of this research work, we follow a typical *design research* approach of a review of current 3D projects (background research), prototype development, and evaluation. This publication presents our prototype and its evaluation. The goal of the project was to develop a prototype which would display financial market data in an immersive 3D environment. Financial market data is often extremely dynamic and readily available in high-frequencies. Information and trading providers such as Bloomberg or Reuters provide live market information to subscribers which is often used to make investment decisions. Recently there has been increased interest surrounding virtual reality and 3D technology in presenting this dynamic data beyond traditional visualization techniques which are often presented in time-series using various charts or variations of heatmaps. Interest in this area has been driven by the needs of industry professionals, in particular traders who often rely on multiple physical trading screens to keep up-to-date with high-frequency data and information as soon as it becomes publicly available. One of the goals of this project was to visualize dynamic data in a manner which allows for easy summarization.

3 Background Research: Review of 3D Projects

Table 1 illustrates a set of different projects and approaches utilizing 2D and 3D means of visualizing of financial data. They utilize different visualization technologies, interaction modalities, and 3D features.

In principle we can characterize current approaches into (1) 3D visualizations moving traditional visualization techniques, such as heatmaps, charts, or scatterplots into the 3D space; and (2) projects providing innovative attempts and providing new solution towards the value that 3D allows. Especially the latter category of projects is rather fascinating, as these projects make use of a new media technology to present data allowing viewers another angle and perspective on data. Adding new dimensions, spatial relationships, geographic meaning, and creating more appealing and easier to understand visualization provide many new opportunities for understanding and interpreting the underlying data. However, one of the main questions remains – what added value does 3D give, and where is 3D useful in the context of financial visualization.

Table 1. Overview of various projects related to 3D visualizations in financial industries.

Project	Description	3D features	Interaction	Technology	Ref.
Thomson Reuters Eikon	Traditional trading desk with some standard 3D features	Combining heatmaps and 3D plotting	Traditional interaction methods	2D screens	[7]
Citi Holographic Workstation	Augmented reality workstation for financial trading	AR, combination of 2D elements and AR	Voice, gesture, keyboard	HoloLens and 2D screens (AR+2D)	[8]
Comarch	Trading desks with a futuristic 'game' alike look for wealth management	Cross-media visualization of banking data		2D screen	[9, 10]
QuantVR	City environment with streets as metaphor for visualization of the stock market	Tree maps modeled in 3D and advanced visualizations of trades as comets, ticker tubes, etc	Typical 3D navigation and interaction modalities	Displaying of 3D scenes on screen	[11]
InVizble	3D environment emphasizing of typical market data in 3D	Very advanced 3D representations of Kepler views, grid array Views, 3D scatter plots, tube views	Typical 3D navigation and interaction modalities	Display of 3D scenes on screens, also enabling head mounted displays or other modalities	[12]
ScienceGL	Scientific tool to visualize stock information in three dimensions	Information and science centered tool to visualize heatmaps, line graphs, etc. in 3D	Simple interaction modalities related to interacting with charts	Representation of data in 3D	[13]

(continued)

Table 1. (continued)

Project	Description	3D features	Interaction	Technology	Ref.
Stock City (Fidelity Labs)	City representation of investor portfolios	3D city models representing stocks, prices, and information about stocks	Typical 3D navigation and interaction modalities	Display of 3D scenes on 2D screens	[14]
VRPPortunity (Salesforce)	Creative approach towards visualizing stock information	Utilization of physical dynamics, color and animations to illustrate information	Advanced interaction and navigation modalities	Display of 3D, and 2.5D models to visualize information	[15]
Bloomberg Terminal	Arrangement of 2D display in an 3D environment	Utilization of 3D to make use of the limited screen space through arranging virtual 2D monitors in 3D	Simple 3D navigation and interaction models	Simple 3D representation of 2D screens	[16, 17]
DXFeed	Representation of technical stock data through AR in an office space	Similar to other tools offering 3D functionality to represent 2D information	Interaction based on AR	Display of stock data in an augmented office space	[18]
Nirvaniq Labs	Metaphors for representing stock information through maps, 3D charts, 3D line charts, and other advanced modalities	Very advanced re-thought representations of typical stock exchange tools as charts, line charts, or maps. Sophisticating use of new representations.	Efficient utilization of 3D navigation and interaction modalities	Typical 3D technology, however, the strength of the project is the advanced representation through enhancing traditional 2D representation	[19]
Looker	Utilization of HMDs to explore financial data	Exploration of traditional charts in a 3D space	Typical HMD based interaction modalities and interaction devices	HMD based environment	[20]

4 Prototype Development of ElectrAus

The goal of the prototype was to illustrate an immersive, interactive 3D visualization on large scale 3D displays, considering the following variables as part of the visualization:

- average price/day based on price given in defined intervals;
- ratio of renewable energy to traditional energy generation e.g. coal powered generated electricity;

- demand and supply of electricity over time at different frequencies;
- scheduled demand and supply based on different regions;
- net electricity import between interconnected regions.

Settlement Date	Dispatch Price ($/MWh)	Scheduled Demand (MW)	Scheduled Generation (MW)	Semi Scheduled Generation (MW)	Net Import (MW)	Type
09/02/2018 00:00	68.99	7977.25	7643.344	100.796	-199.84	Actual
09/02/2018 00:05	67.11332	7920.76	7614.528	99.202	-169.39	Actual
09/02/2018 00:10	59.04981	7821.55	7469.15	96.82	-231.5	Actual
09/02/2018 00:15	84.21912	7855.74	7515.505	96.485	-197.89	Actual
09/02/2018 00:20	77.3176	7827.72	7422.253	93.977	-270.84	Actual
09/02/2018 00:25	68.98995	7806.98	7425.212	90.098	-249.17	Actual
09/02/2018 00:30	68.55261	7697.29	7256.23	90.9	-307.21	Actual
09/02/2018 00:35	68.99	7710.7	7178.887	84.083	-409.53	Actual
09/02/2018 00:40	76.19568	7756.77	7160.862	77.328	-468.7	Actual
09/02/2018 00:45	73.81824	7705.78	7151.415	73.525	-439.07	Actual
09/02/2018 00:50	68.99	7602.13	7053.973	72.097	-435.1	Actual
09/02/2018 00:55	69.0773	7589.23	7063.58	67.46	-418.22	Actual
09/02/2018 01:00	69.07602	7527.85	7039.043	60.007	-387.47	Actual
09/02/2018 01:05	69.0585	7343.1	6856.633	56.807	-389.53	Actual
09/02/2018 01:10	70.0895	7370.88	6798.114	55.826	-469.94	Actual
09/02/2018 01:15	69.04793	7224.25	6731.867	50.953	-402.55	Actual
09/02/2018 01:20	74.42	7312.87	6715.222	49.138	-500.34	Actual
09/02/2018 01:25	70.09038	7283.93	6715.695	48.705	-474.35	Actual
09/02/2018 01:30	69.12643	7228.32	6664.251	46.289	-479.17	Actual
09/02/2018 01:35	87.00001	7207.56	6674.551	46.199	-439.59	Actual
09/02/2018 01:40	81.44933	7227.62	6551.211	45.399	-580.38	Actual
09/02/2018 01:45	76.00119	7169.81	6399.086	41.774	-678.04	Actual

Fig. 2. Example input data for the prototype.

4.1 Australian Energy Market Data

Data used for the project was obtained from publicly available data published by the *Australian Energy Market Operator (AEMO)*.[1] There are two main distinct geographically located and administered energy markets for which data is readily available from AEMO. The first is the *National Electricity Market (NEM)* which connects the following States: Queensland, New South Wales (including the Australian Capital Territory), Victoria, South Australia, and Tasmania. The other large market for which data is readily available is the *Wholesale Electricity Market (WEM)* in the *South West Interconnected System (SWIS)* which applies only to Western Australia.

The electricity market in Australia is well suited as a starting point to data visualization, as data is provided in 5-min or 30-min intervals[2]. The proof of concept was built around a 1-day time period in 5-min intervals for the different states' electricity markets in Australia. However, the interval can be extended to longer time periods (by aggregating or finding an average price) with the potential to summarize and compare multiple markets and multiple variables. The ticks or interval times for data are frequent enough to present meaningful information for a single day which can easily be extended to information of shorter intervals (e.g. millisecond intervals for some data such as foreign exchange rates in well traded currencies).

---

[1] Available from: https://www.aemo.com.au/.

[2] Data is available in 5-min intervals for the National Electricity Market and 30-min intervals for the South West Interconnected System.

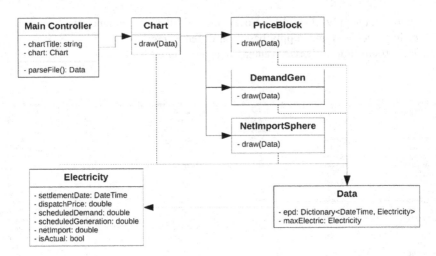

Fig. 3. Class diagram of the overall software modules.

4.2 Data and Variables

The project presents visualized information of the following different electricity market variables which allows for high level summarization of numerically-based data. For the purposes of this project some of the intricate market microstructure is aggregated (see Fig. 2):

- **Dispatch Price:** price determined in 5-min intervals for the NEM, which are averaged to spot prices over a 30-min interval;
- **Scheduled Demand:** electrical power requirement (in megawatts, or MW) which includes scheduled generation, interconnected import/exports including losses and regional scheduled loads;
- **Scheduled Generation:** energy generated and sent out by all scheduled generators (in megawatts, or MW);
- **Net Import:** the net power imported/exported by the relevant region at its inter-regional boundaries;

Figure 2 presents a snapshot of the information that was used in the project. The information comprises of 5-min intervals or settlement dates.

4.3 System Architecture

The system has been developed using the Unity 3D game engine [21]. Besides the components from Unity, we developed a CSV file parsing module, simple data analytics, a 3D charting library, and created 3D models for the prototype. A basic class diagram is illustrated in Fig. 3.

4.4 Design Considerations

The following design guidelines have been applied in designing the prototype (see Table 2):

Table 2. Overview of designs used in our 3D visualization

Visualization Designs of Various Variables	Variables involved in the Visualization Designs
	Visualization of electricity demand and generation throughout seasons
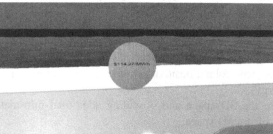	Ratio of renewable energy produced in Australia symbolized as wind turbines
	Average price/day symbolized through a coin
	Scheduled electricity demand and generation

(*Continued*)

Table 2. (*Continued*)

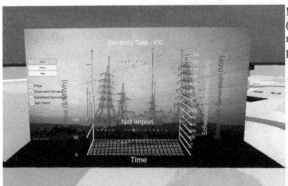

Main chart to display data (dispatch price, scheduled demand and generation, net import)

Electricity net import

- Kevin Lynch's architectural guidelines for city planning (e.g. path, edges, district, nodes, landmarks and focus points) [22];
- Guidelines for color palette creation for visualizing stock market data in a 3D immersive environment (e.g. contrast, visual clues, palette, gradients, …);
- Development of symbols and semiotics of different 3D models to illustrate the essentials of this data (e.g. objects, colors, positive/negative events, ratios, seasonal differences, sustainable energy);
- Migration of 2D charts into the 3D space and providing additional information through efficient utilization of the 3D space;

5   Evaluation and Conclusions

5.1   Prototype Evaluation

One potential measurement on the success of the proposed framework is to evaluate its effectiveness on providing timely and accurate information to decision makers. The variables considered should provide holistic information for suppliers to better manage electricity supply through the day as well as important insight for investors on the difference in demand between different regions. Thus, evaluation can be in the form of feedback from both investors and electricity market operators.

Figure 4 illustrates the current prototype displayed on the Cylinder Display at the Curtin HIVE (Hub for Immersive Visualisation and eResearch). The current prototype represents an initial trial prototype for further investigation and testing.

Fig. 4. Visualization in the Cylinder Display at the Curtin HIVE (Hub for Immersive Visualisation and eResearch).

5.2 Added Value of 3D for Financial Visualization

We concluded, that the main advantages for using 3D, VR, and immersive technologies in finance are:

1. presentation of large amounts of data in a limited space;
2. overcomes space limitations due to a restricted physical space;
3. additional data supporting meaningful additions of information;
4. support of cognitive functions as e.g. pre-attentive processing;
5. easier knowledge and information discovery;
6. visual appealing presentation of qualitative and quantitative data;
7. representing inherently complex data in simplified ways;

Financial market data is inherently complex. For instance, the current application area consists of data from electricity generator i in region j at time t This represents a typical multi-index data with hierarchical structure in the first two indexes. Informative presentations of these data types have always been challenging, especially when the presentations are restricted to 2D format.

3D visualization allows the presentation of large amounts of data in a limited space environment. It also allows effective combinations of both quantitative and qualitative data, which provide users with a holistic view of available information. Despite a limited physical available space, users can explore the data from different angles and perspectives, which is not possible in a 2D space representation. The added dimension also allows adding more data to support the presentation data and add meaningful additions to the data. Simple navigation facilities support data exploration, and by adding more complex interactive elements, the engagement process of a user can be increased. Despite the additional dimension requiring a higher cognitive load, it supports pre-attentive processing of the environment by humans [23]. The use of a virtual 3D environment also allows separation of different sets of data in an intuitive way. However, for the user of the system, quantitative data is presented in a form that it is more visually appealing. This is a clear advantage e.g. if the system is designed for end-users without a solid expertise in financial data. For professionals, e.g. regulators, a data presentation in 3D eases the process of discovering irregularities. Professionals benefit from additional dimensions allowing, data to be presented in a limited physical space in a way that supports cognition and makes the data easier to understand.

5.3   Future Work

We especially would like to investigate how 3D supports (1) the exploration of large amounts of multidimensional financial information and data; (2) data storytelling; (3) extend the current prototype to a fully functional real-time data display in immersive environments; (4) how 3D visualization affects the understanding and cognitive efforts to understand underlying data in the context of *Cognitive Big Data* [24, 25] and (5) adequate intelligent interaction designs (see e.g. [26, 27]) through creative thinking and design thinking [28–30].

Acknowledgements. We would like to thank the Curtin HIVE and its team (namely Andrew Woods) [31], for help in creating the project prototype. We also would like to thank the *Curtin Institute for Computation (CIC)* for the financial grant which supported the project development. The project was conducted as part of the 2017/2018 HIVE summer internship scheme [32].

References

1. Bohn, N., Rabhi, F.A., Kundisch, D., Yao, L., Mutter, T.: Towards automated event studies using high frequency news and trading data. In: Rabhi, F.A., Gomber, P. (eds.) FinanceCom 2012. LNBIP, vol. 135, pp. 20–41. Springer, Heidelberg (2013). https://doi.org/10.1007/978-3-642-36219-4_2
2. Rabhi, F., et al.: Large-scale news processing: concepts and challenges. http://www.academia.edu/868361/Large-Scale_News_Processing_Concepts_and_Challenges
3. Lugmayr, A., Gossen, G.: Evaluation of methods and techniques for language based sentiment analysis for DAX30 stock exchange - a first concept of a "LUGO" sentiment indicator. In: Lugmayr, A., Risse, T., Stockleben, B., Kaario, J., Pogorelc, B., Asensio, E.S. (eds.) Proceedings of the 5th International Workshop on Semantic Ambient Media Experience (SAME) - in Conjunction with Pervasive 2012. Tampere University of Technology (TUT), Newcastle (2012)
4. Robertson, C.S., Rabhi, F.A., Peat, M.: A service-oriented approach towards real time financial news analysis. In: Lin, A., Foster, J., Scifleet, P. (eds.) Consumer Information Systems and Relationship Management: Design, Implementation, and Use. IGI Global, Hershey (2013)

5. Lugmayr, A.: Predicting the future of investor sentiment with social media in stock exchange investments: a basic framework for the DAX performance index. In: Friedrichsen, M., Muehl-Benninghaus, W. (eds.) Handbook of Social Media Management. Media Business and Innovation, pp. 565–589. Springer, Berlin Heidelberg (2013). https://doi.org/10.1007/978-3-642-28897-5_33

6. Lugmayr, A., Sutinen, E., Suhonen, J., Sedano, C.I., Hlavacs, H., Montero, C.S.: Serious storytelling – a first definition and review. Multimed. Tools Appl. **76**(14), 15707–15733 (2017). https://doi.org/10.1007/s11042-016-3865-5. ISSN: 1573-7721

7. Reuters, T.: Thomson Reuters Eikon. https://financial.thomsonreuters.com/en/products/tools-applications/trading-investment-tools/eikon-trading-software.html

8. 8ninths: Cigi Holographic Workstation. http://8ninths.com/case-study/citi-holographic-workstation/

9. Comarch: https://www.comarch.com/finance/

10. Comarch: Virtual reality meets financial services (Comarch at Finovate 2016). https://www.youtube.com/watch?v=kpacQAhJIVY

11. QuantVR: Quant VR (2015). https://www.techrepublic.com/article/quantvr-wants-to-turn-stock-market-data-into-immersive-virtual-reality-experiences/

12. In-Vizble: BLOG: videos and discussion on design and development with the HoloLens. http://www.in-vizible.com/videoblog.html, https://www.youtube.com/c/MichaelPeters-In-Vizible

13. ScienceGL: 3D scientific visualization for stock market analytics. new tools for stock market trading (2013). http://www.sciencegl.com/help_stock/stock_market_3d.html

14. FidelityLabs: Fidelity Labs: StockCity for Oculus Rift (2014). https://www.youtube.com/watch?v=YQ2-8_2Vwpw

15. Salesforce: VRpportunity. https://github.com/CodeScience/VRpportunity

16. Bloomberg: Bloomberg Terminal. https://www.bloomberg.com/professional/solution/bloomberg-terminal/

17. Seward, Z.: Virtual reality headset oculus rift meets the Bloomberg terminal. https://qz.com/218129/virtual-reality-headset-oculus-rift-meets-the-bloomberg-terminal/

18. dxFeed: dxFeed Market Data - VR/AR (2017). http://www.dxfeed.com/vrar

19. Labs, nirvaniq: https://nirvaniq.com/

20. Looker: https://looker.com/solutions/fintech

21. Unity: https://unity3d.com/

22. Lynch, K.: The Image of the City. MIT Press, Cambridge (1960)

23. Sylvester, B.: The visualization of financial data: a review of information visualization tools in the financial data domain (2008). https://www.slideserve.com/more/the-visualization-of-financial-data

24. Lugmayr, A., Stockleben, B., Scheib, C., Mailaparampil, M.: Cognitive big data. survey and review on big data research and its implications: what is really "new"? cognitive big data! J. Knowl. Manag. (JMM) **21**(1), 197–212 (2017)

25. Lugmayr, A., Stockleben, B., Scheib, C., Mailaparampil, M., Mesia, N., Ranta, H.: A comprehensive survey on big data research and it's implications - what is really 'new' in big data? it's cognitive big Data. In: Ting-Peng, L., Shin-Yuan, S.-I.C., Patrick, H., Chau, Y.K. (ed.) Proceedings of the 20th Pacific-Asian Conference on Information Systems (PACIS 2016) (2016)

26. Pogorelc, B., et al.: Semantic ambient media: from ambient advertising to ambient-assisted living. Multimed. Tools Appl. **58**, 399–425 (2012)

27. Lugmayr, A., Risse, T., Stockleben, B., Kaario, J., Laurila, K.: Special issue on semantic ambient media experiences. Multimedia Tools Appl. **44**, 331–335 (2009)

28. Lugmayr, A.: Applying "design thinking" as a method for teaching in media education. In: Proceedings of the 15th International Academic MindTrek Conference: Envisioning Future Media Environments, pp. 332–334. ACM, New York (2011)

29. Lugmayr, A., Jalonen, M., Zou, Y., Libin, L., Anzenhofer, S.: Design thinking in media management education - a practical hands-on approach. In: Lugmayr, A., Risse, T., Stockleben, B., Kaario, J., Pogorelc, B., Asensio, E.S. (eds.) 4th Semantic Ambient Media Experience (SAME) Workshop in Conjunction with the 5th International Convergence on Communities and Technologies. Tampere Univ. of Technology (TUT), Brisbane, Australia (2011)

30. Lugmayr, A., Stockleben, B., Zou, Y., Anzenhofer, S., Jalonen, M.: Applying design thinking in the context of media management education. Multimed. Tools Appl. **71**(1), 1–39 (2013)

31. Woods, A., Datta, S., Bourke, P., Hollick, J.: The design, install and operation of a multi? user, multidisplay visualisation facility (in preparation)

32. Lim, Y.J., Lugmayr, A.: Interactive Creative Visualisation of Financial Markets Data in the Energy Sector. Curtin University (2018)

Document Representation for Text Analytics in Finance

Jan Roeder[(⊠)] and Matthias Palmer

Chair of Electronic Finance and Digital Markets, University of Goettingen,
Goettingen, Germany
{jan.roeder,matthias.palmer}@uni-goettingen.de

Abstract. The automated analysis of unstructured data that is directly or indirectly relevant to developments on financial markets has attracted attention from researchers and practitioners alike. Recent advances in natural language processing enable a richer representation of textual data with respect to semantical and syntactical characteristics. Specifically, distributed representations of words and documents, commonly referred to as embeddings, are a promising alternative. Consequently, this paper investigates the utilization of these approaches for text analytics in finance. To this end, we synthesize traditional and more recent text representation techniques into a coherent framework and provide explanations of the illustrated methods. Building on this distinction, we systematically analyze the hitherto usage of these methods in the financial domain. The results indicate a surprisingly rare application of the outlined techniques. It is precisely for this reason that this paper aims to connect both finance and natural language processing research and might therefore be helpful in applying new methods at the intersection of the respective research areas.

Keywords: Document representation · Text mining · Word embeddings · Conceptual framework · Literature review

1 Introduction

The digitization of social and economic transactions allows individuals and organizations to capture large amounts of structured and unstructured data [1]. While that was already possible at a macro level, micro level interactions on social media or between businesses are currently also recorded and available for analysis [2]. Researchers and practitioners aiming to create insights from textual data are confronted with large corpora. In general, text mining techniques are long established and have been successfully applied to various areas ranging from marketing and sociology to finance and accounting. The financial domain lends itself to this type of analysis because relevant documents such as news articles, social media posts or earnings conference calls and the accompanying analyst reports are readily available [3]. Automated text analysis has been applied to a variety of challenges in the financial domain. Examples include the prediction of exchange rates [4] and stock prices based on financial news [5] or social media data [6]. Typical tasks of text mining are document summarization,

© Springer Nature Switzerland AG 2019
N. Mehandjiev and B. Saadouni (Eds.): FinanceCom 2018, LNBIP 345, pp. 131–145, 2019.
https://doi.org/10.1007/978-3-030-19037-8_9

classification, and clustering. They are commonly performed with machine learning (ML) algorithms [7]. Determinants of the successful application of an ML algorithm are not only the choice of the algorithm itself or the amount of available data but also the way the data is represented, i.e., how the text is captured in a machine-readable form [8]. Traditionally, linear ML models are trained with high-dimensional and sparse vectors since the most common approach to describe a word numerically is one-hot encoding [9]. Here, a vocabulary of unique words contained in a text corpus is created and a single word is represented by an index in that vocabulary. This is also known as a local representation, since every entity to be represented is assigned its own dimension in the vector space [10]. However, this approach comes with several drawbacks, such as data sparsity and no concept of semantic similarity between words [11, 12]. To address these shortcomings, there has been a shift to a lower-dimensional, dense representation of words, commonly referred to as word embeddings [9]. These word embeddings take the form of a distributed representation. In contrast to a local representation, a word is not represented by a distinct dimension but through the combination of activity in many different dimensions [13]. This is accompanied by advantages such as a lower dimensionality and an improved ability to generalize [13]. Another reason for representing words with real-valued and lower-dimensional vectors is that better performance can be achieved by leveraging local smoothness properties of the classification algorithms [8, 14, 15]. The concept of a distributed representation of words can be further generalized to text sequences of variable length, e.g., paragraphs or documents [16]. The sparsity and volume of unstructured data are identified as an avenue for future research in the context of an information systems research agenda [17]. Both issues are at least partially addressed by the distributed representation of textual artifacts.

While embedding approaches have shown superior performance in tasks like sentiment analysis or information retrieval [16, 18], there seems to be a lack of research that evaluates the potential of these techniques for applications in finance. A recent literature review on research utilizing text mining for financial market prediction shows that the relatively simplistic bag-of-words model is applied in about three quarters of analyses [19]. While it is expected that document embeddings are not necessarily superior in general, it would be valuable to explore use cases where they show an improved effectiveness. Several literature reviews provide comprehensive information on specific areas of text mining such as sentiment analysis [20–22] or topic mining [23, 24], the latter being a specific type of document representation. However, these reviews are not specific to finance. A growing amount of papers, especially in the natural language processing (NLP) research stream, contributes to research by comparing methods from both approaches to specific problems [25] and by giving an overview on different techniques for text mining [21, 26–28] and data representation learning [29]. Additionally, specific application fields, such as social networks [30] or accounting and finance [3], are addressed. Nevertheless, these contributions do not provide an integrative view with respect to recent advances in NLP.

The objective of this paper is to provide a synopsis of the existent work in the financial domain that uses distributed document representations. We aim to identify research gaps and promising avenues for future research. We are not aware of any framework that integrates recent and traditional approaches of text representation in a suitable manner. Thus, we consider the development of such a framework to be an important basis for the following analysis and a helpful tool for text analytics in finance. Relating the document representation techniques to each other might invite researchers to evaluate the benefits of these representations in their applied work. We address the following research question: *In which areas of finance research are distributed document representations already being used and what are the hitherto findings?*

This paper is structured as follows. Section 2 provides the theoretical foundation of document representation. Section 3 describes the research design of the structured literature review. Section 4 presents the results of the literature review. Section 5 discusses implications and limitations. Section 6 concludes the paper and gives an outlook on future research directions.

2   Theoretical Foundation

2.1   Text Mining Process

The process of text mining or knowledge discovery from text (KDT) [31] refers to the process of knowledge discovery from databases (KDD) [32]. Various manifestations of the text mining process can be found within the literature. Some deal with high-level architecture [33–35] and others are focused on specific application areas, such as sentiment analysis [21], market prediction [19], social networks [30], big data analysis [36], and patent analysis [37]. Selected publications have paid attention to pre-processing tasks [35, 38–40]. However, we are not aware of a conceptual overview that focuses on document representation techniques and the resulting representations. Based on the text mining processes used in the mentioned literature and in line with the KDD process, we propose a text mining process that acts as a parent framework as shown in Fig. 1. We split the transformation step of the KDD process into two sub-steps: feature extraction and feature representation. This is done to disentangle the concerns of transforming text to a numeric representation and consequently manipulating this representation by feature weighting or dimensionality reduction techniques. The proposed text mining process consists of the following steps:

- (1) *Data selection*: selecting textual sources, e.g., news, social media, regulatory filings, or financial statements and if necessary, choosing a subset of the data
- (2) *Text pre-processing*: conducting morphological, syntactical, and semantic analysis, e.g., tokenization, sentence boundary detection, transformation to lower case, stop word removal, removal of punctuation, whitespace stripping, removing numbers, stemming, lemmatization, and substitution of hyponyms and synonyms
- (3) *Feature extraction*: transforming textual artifacts into numerical features
- (4) *Feature representation*: weighting features and reducing dimensionality for creating a meaningful representation
- (5) *Data analysis*: classification, clustering, association analysis, and predictive modeling, e.g., for sentiment analysis or document similarity analysis
- (6) *Interpretation and evaluation*: visualizing data or calculating evaluation metrics, e.g., accuracy, precision, recall, F-measure, and relevance

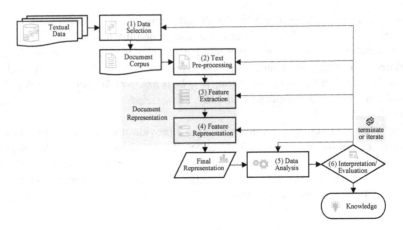

Fig. 1. Text mining process with a focus on document representation

2.2   Conceptual Document Representation Framework

The anchor point of our framework is the structure of the final document representation, as shown in Table 1. We distinguish between sparse and dense representations. The framework explicates the steps feature extraction and feature representation of the text mining process in Fig. 1 in detail. The subdivision coincides with the steps in [41], which are applied in a similar context. For feature extraction, a general division can be achieved by differentiating between the inputs to the representation technique, namely count-based and prediction-based inputs [42]. Feature representation is subdivided into feature weighting, also known as term weighting [43, 44], and dimensionality reduction [45, 46]. In general, dimensionality reduction is performed by feature selection or feature extraction methods. For text mining, a more granular division is better suited from our perspective. Therefore, we split dimensionality reduction into feature selection and feature transformation, also known as feature discovery [47]. In feature selection, a subset of the features is selected. Feature transformation generates a new feature set and creates a new, lower dimensional feature space while retaining as much information as possible, resulting in a dense representation. Here, a further distinction is made between the method used, the type of representation created, and the final structure of this representation. Furthermore, it must be distinguished between three levels of text representation: word, document, and document collection. Based on the subdivision between bag-of-words and word embedding approaches, we discuss the individual levels in the next section and compare between the two approaches. We do not discuss the document and document collection levels separately, as no further implications can be derived from this distinction. This framework is not intended to contain all conceivable steps and methods for document representation. Rather, it should serve as a fundament to the literature analysis as it allows to situate the identified literature in the overall text mining context.

Table 1. Conceptual document representation framework: The steps are printed in bold, specific methods are in italics, the type of representation is in normal letters, and the final structure of the representation is underlined. The methods mentioned are exemplary and not exhaustive.

SPARSE REPRESENTATION	DENSE REPRESENTATION		
3 FEATURE EXTRACTION			
3.1 Word level			
One-hot encoding	*Word2Vec*	*GloVe*	
Full word representation (count-based)	Word embedding (prediction-based)		
<u>Sparse representation</u>	<u>Dense representation</u>		
3.2 Document level			
Addition of one-hot vectors	*Addition of word vectors*	*Doc2Vec*	*Autoencoder*
Bag-of-word vector	Document embedding		
<u>Sparse representation</u>	<u>Dense representation</u>		
3.3 Document collection level			
Concatenation of bag-of-word vectors	*Concatenation of document embedding*		
Term-document matrix	Matrix of document embeddings		
<u>Sparse representation</u>	<u>Dense representation</u>		
4 FEATURE REPRESENTATION			
4.1 Feature weighting			
Binary / *TF* / *IDF* / *TF-IDF*			
Term-document matrix (weighted)			
<u>Sparse representation</u>			
4.2 Dimensionality reduction			
4.2.1 Feature selection methods	**4.2.2 Feature transformation methods**		
No selection / *Feature scoring*	*LSA/LSI*	*pLSA* / *LDA*	*Autoencoder*
Term-document matrix (weighted)	Document matrix (concept-centric)	Document matrix (topic-centric)	Matrix of document embeddings
<u>Sparse representation</u>	<u>Dense representation</u>		

Our framework does not focus on N-grams, which is an approach that considers word groups to better illustrate semantic contexts, and on end-to-end text mining approaches. End-to-end approaches are predominantly based on deep neural networks that accept a series of words encoded as word embeddings as an input, implicitly create an internal document representation, and learn to perform classification for tasks like sentiment analysis or topic classification [48, 49]. This stands in contrast to traditional ML models, which are usually reliant on an intermediate document representation with fixed dimensionality. We argue that these representations are accessible for a wider audience and provide a middle ground between complexity and potential performance improvements. Hence, we consider end-to-end approaches to be out of scope for this study but highly recommend an analysis regarding their potentials and possible applications for text mining in finance.

2.3 Word Level Representation

The richness of a document representation is constrained by the way individual words are represented and consequently by the semantical and syntactical information contained in this representation; especially, since any information that is discarded when transforming words into a more machine-readable form cannot be restored at a later point in time. The goal of feature extraction at the word level is to achieve a mathematical representation of every word contained in the text to prepare the data for classification or regression with ML algorithms. This is achieved using methods that create vectors that aim to capture semantic and syntactic regularities. Most of the literature on text mining for financial forecasting uses one-hot encoding with the dimensionality corresponding to the number of unique words in the corpus. In other words, for every word in the text, a vector is created that has the length of the vocabulary of the document corpus with the dimension corresponding to the index of the specific word being one and all other dimensions being zero. Here, the word order is irrelevant and the words are mutually independent. For model training, sparsity is a threat since it leads to a poor estimate for model parameters of rare words [12]. Also, individual words that are not contained in the training corpus cannot be used in a later model application [12].

The limitations of one-hot word representations have prompted researchers to investigate other methods such as word embeddings [15, 50]. They predict surrounding words and are therefore referred to as predictive models, which have been shown to be superior to one-hot encoding [51]. Word embeddings are based on the assumption of the distributional hypothesis, which states that the meaning of a word can be deduced from the context words that it frequently co-occurs with [52]. This should not be confused with the distributed representation, the low-dimensional and real-valued form that these word embeddings take. The latent dimensions of word embeddings, which usually range from 100 to 300, aim to encode syntactic and semantic regularities of each word. This comes with multiple advantages. It reduces the computational effort since the dimensionality of the word vectors is not determined by the vocabulary size of the corpus [9]. Furthermore, second-order co-occurrences, also called paradigmatic parallels [53], are encoded in the word vector. This means that "father" and "dad" have a similarity (typically measured with the cosine similarity) of close to one. This stands in contrast to the one-hot encoded model, where the words would exhibit a cosine similarity of zero. Thus, similar words are assigned a similar representation, which improves generalization power [9]. A fundamental aspect of word embeddings is the fact that they can be inferred from a large document corpus without requiring labeled data. The goal is to learn a representation for the unlabeled data, which in turn can lead to drastic performance improvements when used for downstream tasks such as classification [12, 54]. Generally, dense word representations can be induced using matrix factorization [55], context-based prediction models (Word2Vec) [11] or even by leveraging the advantages of both approaches through a hybrid approach (Glove) [56].

Although it has been shown that prediction-based approaches are not necessarily superior [57], we will put emphasis on these comparatively young methods.

2.4   Document Level Representation

Typically, classifiers in text mining applications are trained at the document and document collection level, where each document represents a single data point. Therefore, the document and document collection levels can be considered together. To create a document representation, the constituent word vectors of a document are aggregated. In the case of the bag-of-words approach, the aggregation of document vectors results in a vector space model that allows for analyses on the corpus level [58]. The resulting representation has the same weaknesses as the one-hot representation of individual words. Furthermore, the representation is high-dimensional since its dimensionality is determined by the number of unique words in the corpus or is limited explicitly. However, this issue is largely mitigated by algorithms, e.g., support vector machines (SVM), that can take advantage of sparsity [9].

Following the feature extraction, various techniques can be applied in order to lower the number of irrelevant features in the dataset. In the framework in Table 1, these steps are subsumed under feature representation. The basic step in this part is feature weighting. Binary and term frequency (*tf*) weighting are applied on the document level. Inverse document frequency (*idf*) assumes that the importance of a word is inversely related to its frequency in a document collection. *Tf-idf* is a combination of the latter two, increasing with a higher number of word appearances and at the same time weighting this by the word frequency in the document collection to reduce the influence of more common words. One step of the following dimensionality reduction is feature selection by methods such as information gain, mutual information, and χ^2 statistics [59]. However, these representations are still sparse and high-dimensional.

The alternative step of feature transformation offers a solution by summarizing methods that form a dense representation from a vector space model. One way is to discover latent structures in a document collection. Latent semantic analysis (LSA) is a method that decomposes the initial representation and projects the most relevant features onto a subspace [55]. That is, it applies singular value decomposition (SVD) on a term-document matrix with the result that topics are assumed to be orthogonal. Even though this assumption does not seem to be very realistic, it works well in practice, comparable to principal component analysis (PCA), which is used in regression problems to reduce the number of features. On this basis, models were developed that assume documents to be a composition of various topics, e.g., probabilistic latent semantic analysis (pLSA). These topics are assumed to be a distribution of the respective vocabulary of a document corpus and allow topics to be non-orthogonal. Latent Dirichlet allocation (LDA) is a model that extends pLSA by applying Dirichlet distributed priors to the document-topic and topic-word distributions leading to better results by preventing overfitting [60]. Nonetheless, research also suggests that LSA is not necessarily inferior to the later developed models [61]. These are some of the most widely known models that should give an intuition of how representations have advanced over time. Another way to create a representation is to take advantage of techniques based on neural networks, e.g., autoencoders [62], neural topic models [63], and restricted Boltzmann machines [64]. These techniques aim to create encodings that correspond to a latent representation, similar to topic models [65].

As described earlier, to translate the textual artifacts into a representation that can be used as an input for traditional ML algorithms, such as SVMs that show good performance for text mining tasks [66], the representation by fixed-length vectors is necessary. Even though it would be possible to aggregate the word vectors contained in a document by calculating an average, local semantic and syntactic regularities would be lost in the process. Thus, more sophisticated techniques are required to aggregate the constituent word vectors of a document. One of the more prominent algorithms to achieve this task is the paragraph vector. The method incorporates the word order, outputs a fixed length document representation, and uses the semantic expressiveness of word embeddings [16]. It has been shown that paragraph vectors exhibit superior performance in tasks such as information retrieval and sentiment analysis, compared to representations using a simple bag-of-words or LDA representation [16].

3   Research Methodology

The outlined framework forms the basis for the following literature analysis that builds upon the methodological work of [67, 68]. These papers provide an overall processual guideline, while the developed framework allows to situate the identified literature at the content level. In general, the structured literature analysis identifies existing research in a specific domain [69] and highlights research gaps to identify avenues for future research [70]. The literature analysis should result in a concept matrix, thereby ensuring a concept-centric perspective [68]. We divide the search process into two phases: First, an outlet for the search is defined and an initial search is performed. Second, a backward search uses the references in the initial articles to identify prior research articles. To ensure reproducible results, the following data should be provided: names of the databases, search strings, criteria for the search process, search date, number of hits per search phase, and the resulting literature [70].

We search for articles in a broad selection of databases to cover as many finance-specific articles as possible. The specific database names and details of the search results are described below. We searched for the following terms in combination with an AND "finance" search term: "word embedding*" OR "document embedding*" OR "document representation*" OR "word co-occurrence*" OR "Word2Vec" OR "Doc2Vec" OR "GloVe" OR "fastText" OR "text embedding*" OR "prediction-based representation*." We excluded textbooks and dissertations from our search, whereas we intentionally include conference articles and working papers, since we assume that applications of the young concept of pre-trained word embeddings, as introduced by [11], have not yet found their way into journals on a large scale. The search results are restricted to the period of 2003 to 2018, since we do not expect to find any relevant results before the first proposition of neural word embeddings by Bengio et al. [15]. This study puts an emphasis on applied work, i.e. the application of distributed document representations in the financial domain.

We derive classification dimensions for the literature search from an article that discusses recent advancements in data, text, and media mining [2]. To create a suitable framework for their literature analysis, they derive what they call the *essential components of information mining*: (1) data type, (2) application area, (3) techniques, (4) tasks, and (5) final objective. We refrain from specifying the data type because it will always be text in this analysis. Two more dimensions are added: representation technique and findings. In addition, we do not define the characteristics of the individual dimensions beforehand but elaborate them during the review process. The following procedure ensures a consistent classification of the articles. First, each article is classified by one of the authors. Second, the classified article is verified with the co-author. Third, the article is discussed in cases of disagreement.

The search conducted in May 2018 resulted in 750 hits with 35 potentially relevant articles that were identified by an analysis of titles and abstracts (hits; potentially relevant): Business Source Complete (38; 5), AISel (13; 6), JSTOR (3; 1), SpringerLink (302; 9), Science Direct (141; 9), EconPapers (3; 2), Emerald Insight (17; 0), IEEE Xplore Digital Library (157; 1), and Wiley Online (76; 2). A subsequent in-depth analysis led to two remaining articles from the initial search. The backward search in the second phase added two more articles. A final check of relevance and the specified search criteria resulted in a definitive number of four articles.

4 Findings

The identified literature is shown in Table 2. What stands out is the relatively low number of four papers that use a distributed document representation to solve a downstream task. In all reviewed papers, this task is a classification problem. In each case, the primary objective is the improved representation of textual data to increase the performance of a classifier. Concerns about an improved efficiency enabled by the lower dimensional document representation do not seem to be an issue. Even though we searched for literature dating back to 2003, the earliest identified publication is from 2016. Three out of four papers rely on word embeddings to represent individual words. We hypothesize that the main reason for this is the implementation of the Word2Vec algorithm in mature software packages like gensim [71]. In contrast, Feuerriegel and Fehrer [72] rely on one-hot encoding. Cerchiello et al. [73] and Rönnqvist and Sarlin [74] aggregate the constituent words of the document with the paragraph vector algorithm as described in [75]. It should be noted that Rönnqvist and Sarlin are co-authors of the paper Cerchiello et al. [73]. Yan et al. [76] compute the average of all word vectors in a document. Since the authors did not use a variety of algorithms for document representation, e.g., paragraph vectors or autoencoders, it is difficult to estimate whether a better performance could have been achieved with alternative approaches. Furthermore, Yan et al. [76] do not compare their proposed approach with a less complex representation method, e.g., bag-of-words. This impedes efforts to disentangle what aspects of the improvements can be attributed to the usage of textual data and which improvements are caused by a semantically richer representation. The same point holds true for Cerchiello et al. [73] since they show an increase in the relative usefulness compared to purely textual or numerical prediction but do not compare different document representations as the basis for their classification. Feuerriegel and Fehrer [72] use recursive autoencoders to capture the ordered context of words. Their approach of using one-hot-encoded words as input for the recursive autoencoder does not conform to our proposed conceptualization. However, it is more common to use word embeddings as input as described in Socher et al. [77]. They compare the neural network approach to a random forest classifier trained on documents represented by a *tf-idf* matrix. For stock price movement prediction, they report a relative improvement in the F1-measure of 7.69% compared to the baseline.

Table 2. Results of the structured literature review

Citation	Application area	Techniques	Tasks	Final objective	Representation techniques	Findings
(Cerchiello et al. 2017)	Finance: Bank distress	Fully connected neural network (hidden layer of 50 nodes, binary classification)	Classification	Improve text-based bank distress classifier by considering additional financial data	Distributed Memory Model of Paragraph Vectors (PV-DM) by Le and Mikolov	Combination of textual and numerical data improves performance of bank distress classification
(Yan et al. 2018)	Finance: Social lending	Random forest, gradient boosting, gradient tree boosting	Classification	Enhance social lending recommendation by leveraging the motivation of the participants	Averaging of word embeddings	Classification and usage of text about motivation of lending platform participants can improve lending project recommendation
(Feuerriegel and Fehrer 2016)	Finance: Stock prediction	Final softmax layer, random forest	Classification	Prediction of stock movements following financial disclosures	Recursive autoencoder	Distributed representation shows performance gain compared to local representation
(Rönnqvist and Sarlin 2017)	Finance: Bank distress	Fully connected neural network (hidden layer of 50 nodes, binary classification)	Classification	Predict bank distress using textual news data to perform risk analysis	Distributed Memory Model of Paragraph Vectors (PV-DM) by Le and Mikolov	Classifier based on financial news data allows the prediction of bank distress events and outperforms majority class prediction

5   Discussion

Apart from the results of the literature review, our proposed framework situates the recent progress in the representation of text artifacts in the overall text mining context. This has potential to benefit future research as a general guideline. We highlight that a distinction depending on the structure of the final representation (dense or sparse) and on the extraction of individual word features (count-based or prediction-based) results in a differentiation of various document representation techniques. This is helpful from different perspectives. First, it can help to structure document representation tasks in other research projects. Second, it allows a fast look-up for which method results in which kind of representation. Furthermore, the connection of NLP research and finance has the potential to sensitize finance researchers for paying attention to NLP research.

As addressed in the research question, the literature review provides an overview of the use of document embeddings in the past years in finance research and reveals that very few studies have applied the described techniques. However, this might have the following two positive effects. First, researchers who have mainly relied on a local representation of text may evaluate embeddings for research projects. Second, document embeddings might be considered a useful alternative by a broader group of researchers in the finance domain. In addition, this work gives practitioners an integrated view of word and document level representation approaches and might be a basis to decide on the sensible application of these techniques.

Even though word and document embeddings are important concepts, this paper chooses to give a general intuition but no guidance on how to use these methods. Also, from our perspective, the framework names the most relevant methods but does not provide a complete list. Within the framework, we have not explicitly referred to the research of Bengio et al. [15] and Collobert and Weston [50]. Although these articles are pioneering for word embeddings, implementations such as Word2Vec and GloVe are easier to apply and therefore more suitable as introductory methods for text mining. We do not want to overlook the fact that document representations based on word embeddings are not interpretable caused by their distributed structure. In contrast, topical representations are easier to interpret, and above all they are an intuitive method for researchers who have no background in NLP. Furthermore, our literature review is limited to the finance domain. We think that a focus on a specific research domain is advisable to ensure a proper classification of the resulting articles, but this approach might miss interesting articles from other domains that apply word embeddings. Moreover, word embeddings do not necessarily have to be used as inputs to create dense embeddings that are fed into classical ML algorithms. Word embeddings might show their full potential as input for deep neural networks to identify position invariant local clues such that the word order is modeled. Thus, future research might analyze specifically if and how those techniques have been used in finance research. Future research might also empirically investigate the advantages of utilizing distributed representations for the analysis of specific finance-related text corpora.

6   Conclusion

Commonly, text mining approaches are based on high-dimensional and sparse vectors for document representation. However, more recent NLP research addresses these shortcomings. In this regard, distributed representations of text such as word and document embeddings are a promising tool. That is why this paper provides a synopsis of existent work that builds on the outlined techniques to create distributed representations of documents in applied finance research. To prepare the literature review, we have connected traditional text mining approaches with word and document embeddings in an overarching framework. This allows for a comparison of the respective approaches. The subsequent structured literature overview shows that little research on the actual application of the outlined methods has been conducted so far. Although distributed document representations based on word embeddings are not necessarily advantageous for every kind of application, this small number of results is surprising since we are not aware of research that fundamentally refutes the effectiveness of distributed representations in the field of applied finance research. On one hand, the proposed framework can be used by finance researchers as an overview on fundamental distinctions in the field of document representations. On the other hand, it might sensitize finance researchers to the advances in NLP. Considering the results of our analysis, we see applied research on possible applications of distributed document representations as a promising direction for future research in finance.

References

1. Agarwal, R., Dhar, V.: Big data, data science, and analytics: the opportunity and challenge for IS research. Inf. Syst. Res. **25**(3), 443–448 (2014)
2. Gopal, R., Marsden, J.R., Vanthienen, J.: Information mining—reflections on recent advancements and the road ahead in data, text, and media mining. Decis. Support Syst. **51**, 727–731 (2011)
3. Loughran, T., McDonald, B.: Textual analysis in accounting and finance: a survey. J. Acc. Res. **54**(4), 1187–1230 (2016)
4. Jin, F., Self, N., Saraf, P., Butler, P., Wang, W., Ramakrishnan, N.: Forex-foreteller: currency trend modeling using news articles. In: 19th ACM International Conference on Knowledge Discovery and Data Mining, pp. 1470–1473. ACM (2013)
5. Tetlock, P.C., Saar-Tsechansky, M., Mackskasssy, S.: More than words: quantifying language to measure firms' fundamentals. J. Finan. **63**(3), 1437–1467 (2008)
6. Bollen, J., Mao, H., Zeng, X.: Twitter mood predicts the stock market. J. Comput. Sci. **2**(1), 1–8 (2011)
7. Hotho, A., Nürnberger, A., Paaß, G.: A brief survey of text mining. In: LDV Forum, pp. 19–62 (2005)
8. Bengio, Y., Courville, A., Vincent, P.: Representation learning: a review and new perspectives. IEEE Trans. Pattern Anal. Mach. Intell. **35**(8), 1798–1828 (2013)
9. Goldberg, Y.: A primer on neural network models for natural language processing. J. Artif. Intell. Res. **57**, 345–420 (2016)
10. Hinton, G.E.: Learning distributed representations of concepts. In: 8th Annual Conference of the Cognitive Science Society, pp. 1–12. Amherst, MA (1986)
11. Mikolov, T., Chen, K., Corrado, G., Dean, J.: Efficient estimation of word representations in vector space. arXiv preprint arXiv:1301.3781 (2013)
12. Turian, J., Ratinov, L., Bengio, Y.: Word representations: a simple and general method for semi-supervised learning. In: 48th Annual Meeting of the Association for Computational Linguistics, pp. 384–394 (2010)
13. Hinton, G.E., McClelland, J., Rumelhart, D.: Distributed representations. In: Parallel Distributed Processing: Explorations in the Microstructure of Cognition, vol. 1, no. (3), pp. 77–109 (1986)
14. Bengio, Y.: Learning deep architectures for AI. Found. Trends® Mach. Learn. **2**(1), 1–127 (2009)
15. Bengio, Y., Ducharme, R., Vincent, P., Jauvin, C.: A neural probabilistic language model. J. Mach. Learn. Res. **3**, 1137–1155 (2003)

16. Le, Q.V., Mikolov, T.: Distributed representations of sentences and documents. In: 31st International Conference on Machine Learning, pp. 1188–1196 (2014)
17. Abbasi, A., Sarker, S., Chiang, R.H.L.: Big data research in information systems: toward an inclusive research agenda. J. Assoc. Inf. Syst. **17**(2), i–xxxii (2016)
18. De Boom, C., Van Canneyt, S., Demeester, T., Dhoedt, B.: Representation learning for very short texts using weighted word embedding aggregation. Pattern Recogn. Lett. **80**, 150–156 (2016)
19. Nassirtoussi, A.K., Aghabozorgi, S., Wah, T.Y., Ngo, D.: Text mining for market prediction: a systematic review. Expert Syst. Appl. **41**(16), 7653–7670 (2014)
20. Feldman, R.: Techniques and applications for sentiment analysis. Commun. ACM **56**(4), 82–89 (2013)
21. Kearney, C., Liu, S.: Textual sentiment in finance: a survey of methods and models. Int. Rev. Financ. Anal. **33**, 171–185 (2014)
22. Pang, B., Lee, L.: Opinion mining and sentiment analysis. Found. Trends® Inf. Retrieval **2**(1–2), 1–135 (2008)
23. Debortoli, S., Müller, O., Junglas, I., vom Brocke, J.: Text mining for information systems researchers: an annotated topic modeling tutorial. Commun. Assoc. Inf. Syst. **39**(7), 110–135 (2016)
24. Eickhoff, M., Neuss, N.: Topic modelling methodology: its use in information systems and other managerial disciplines. In: 25th European Conference on Information Systems, pp. 1327–1347 (2017)
25. Lee, S., Baker, J., Song, J., Wetherbe, J.C.: An empirical comparison of four text mining methods. In: 43rd Hawaii International Conference on System Sciences, pp. 1–10. IEEE (2010)
26. Agrawal, R., Batra, M.: A detailed study on text mining techniques. Int. J. Soft Comput. Eng. **2**(6), 118–121 (2013)
27. Allahyari, M., et al.: A brief survey of text mining: classification, clustering and extraction techniques. arXiv preprint (2017)
28. Gupta, V., Lehal, G.S.: A survey of text mining techniques and applications. J. Emerg. Technol. Web Intell. **1**(1), 60–76 (2009)
29. Zhong, G., Wang, L.-N., Ling, X., Dong, J.: An overview on data representation learning: from traditional feature learning to recent deep learning. J. Finan. Data Sci. **2**(4), 265–278 (2016)
30. Irfan, R., et al.: A survey on text mining in social networks. Knowl. Eng. Rev. **30**(2), 157–170 (2015)
31. Feldman, R., Dagan, I.: Knowledge discovery in textual databases (KDT). In: 1st International Conference on Knowledge Discovery and Data Mining, pp. 112–117 (1995)
32. Fayyad, U., Piatetsky-Shapiro, G., Smyth, P.: From data mining to knowledge discovery in databases. AI Mag. **17**(3), 37–54 (1996)
33. Fan, W., Wallace, L., Rich, S., Zhang, Z.: Tapping the power of text mining. Commun. ACM **49**(9), 76–82 (2006)
34. Feldman, R., Sanger, J.: The Text Mining Handbook: Advanced Approaches in Analyzing Unstructured Data. Cambridge University Press, Cambridge (2007)
35. Miner, G., Elder, J., Hill, T.: Practical Text Mining and Statistical Analysis for Non-Structured Text Data Applications. Academic Press, Cambridge (2012)
36. Pavlopoulou, N., Abushwashi, A., Stahl, F., Scibetta, V.: A text mining framework for big data. Expert Update **17**(1), 1–23 (2017)
37. Tseng, Y.-H., Lin, C.-J., Lin, Y.-I.: Text mining techniques for patent analysis. Inf. Process. Manage. **43**(5), 1216–1247 (2007)

38. Jing, L.-P., Huang, H.-K., Shi, H.-B.: Improved Feature selection approach TFIDF in text mining. In: 1st International Conference on Machine Learning and Cybernetics, pp. 944–946 (2002)
39. Munková, D., Munk, M., Vozár, M.: Data pre-processing evaluation for text mining: transaction/sequence model. Procedia Comput. Sci. **18**, 1198–1207 (2013)
40. Vijayarani, S., Ilamathi, M.J., Nithya, M.: Preprocessing techniques for text mining – an overview. Int. J. Comput. Sci. Commun. Netw. **5**(1), 7–16 (2015)
41. Lowe, W.: Towards a theory of semantic space. In: Annual Meeting of the Cognitive Science Society (2001)
42. Turney, P.D., Pantel, P.: From frequency to meaning: vector space models of semantics. J. Artif. Intell. Res. **37**, 141–188 (2010)
43. Lan, M., Tan, C.L., Su, J., Lu, Y.: Supervised and traditional term weighting methods for automatic text categorization. IEEE Trans. Pattern Anal. Mach. Intell. **31**(4), 721–735 (2009)
44. Salton, G., Buckley, C.: Term-weighting approaches in automatic text retrieval. Inf. Process. Manage. **24**(5), 513–523 (1988)
45. Manning, C.D., Raghavan, P., Schütze, H.: An Introduction to Information Retrieval. Cambridge University Press, Cambridge (2008)
46. Underhill, D.G., McDowell, L.K., Marchette, D.J., Solka, J.L.: Enhancing text analysis via dimensionality reduction. In: IEEE International Conference on Information Reuse and Integration, pp. 348–353 (2007)
47. Pechenizkiy, M., Tsymbal, A., Puuronen, S.: PCA-based feature transformation for classification: issues in medical diagnostics. In: 17th IEEE Symposium on Computer-Based Medical Systems, pp. 535–540 (2004)
48. Kim, Y.: Convolutional neural networks for sentence classification. In: 2014 Conference on Empirical Methods in Natural Language Processing, pp. 1746–1751. Association for Computational Linguistics (2014)
49. Severyn, A., Moschitti, A.: Twitter sentiment analysis with deep convolutional neural networks. In: 38th International ACM SIGIR Conference on Research and Development in Information Retrieval, pp. 959–962. ACM (2015)
50. Collobert, R., Weston, J.: A unified architecture for natural language processing: deep neural networks with multitask learning. In: 25th International Conference on Machine Learning, pp. 160–167. ACM (2008)
51. Baroni, M., Dinu, G., Kruszewski, G.: Don't count, predict! A systematic comparison of context-counting vs. context-predicting semantic vectors. In: 52nd Annual Meeting of the Association for Computational Linguistics, pp. 238–247 (2014)
52. Firth, J.R.: A synopsis of linguistic theory 1930–1955. In: Palmer, F.R. (ed.) Selected Papers of 1952–1959, pp. 168–205, Longmans, London (1957)
53. Schütze, H., Pedersen, J.: A vector model for syntagmatic and paradigmatic relatedness. In: 9th Annual Conference of the UW Centre for the New OED and Text Research, pp. 104–113 (1993)
54. Goodfellow, I., Bengio, Y., Courville, A.: Deep Learning. MIT Press, Cambridge (2016)
55. Deerwester, S., Dumais, S.T., Furnas, G.W., Landauer, T.K., Harshman, R.: Indexing by latent semantic analysis. J. Am. Soc. Inf. Sci. **41**(6), 391–407 (1990)
56. Pennington, J., Socher, R., Manning, C.: GloVe: Global Vectors for Word Representation. In: 2014 Conference on Empirical Methods in Natural Language Processing, pp. 1532–1543 (2014)
57. Levy, O., Goldberg, Y., Dagan, I.: Improving distributional similarity with lessons learned from word embeddings. Trans. Assoc. Comput. Linguist. **3**, 211–225 (2015)
58. Salton, G., Wong, A., Yang, C.-S.: A vector space model for automatic indexing. Commun. ACM **18**(11), 613–620 (1975)

59. Yang, Y., Pedersen, J.O.: A comparative study on feature selection in text categorization. In: 14th International Conference on Machine Learning, pp. 412–420 (1997)
60. Blei, D.M., Ng, A.Y., Jordan, M.I.: Latent Dirichlet allocation. J. Mach. Learn. Res. **3**, 993–1022 (2003)
61. Bergamaschi, S., Po, L.: Comparing LDA and LSA topic models for content-based movie recommendation systems. In: Monfort, V., Krempels, K.-H. (eds.) WEBIST 2014. LNBIP, vol. 226, pp. 247–263. Springer, Cham (2015). https://doi.org/10.1007/978-3-319-27030-2_16
62. Ranzato, M.A., Szummer, M.: Semi-supervised learning of compact document representations with deep networks. In: 25th International Conference on Machine Learning, pp. 792–799. ACM (2008)
63. Cao, Z., Li, S., Liu, Y., Li, W., Ji, H.: A novel neural topic model and its supervised extension. In: 29th AAAI Conference on Artificial Intelligence, pp. 2210–2216 (2015)
64. Larochelle, H., Bengio, Y.: Classification using discriminative restricted Boltzmann machines. In: 25th International Conference on Machine Learning, pp. 536–543. ACM (2008)
65. Wei, C., Luo, S., Ma, X., Ren, H., Zhang, J., Pan, L.: Locally embedding autoencoders: a semi-supervised manifold learning approach of document representation. PLoS ONE **11**(1), e0146672 (2016)
66. Wang, S., Manning, C.: Baselines and bigrams: simple, good sentiment and topic classification. In: 50th Annual Meeting of the Association for Computational Linguistics, pp. 8–14 (2012)
67. Levy, Y., Ellis, T.J.: A systems approach to conduct an effective literature review in support of information systems research. Inf. Sci. J. **9**, 181–212 (2006)
68. Webster, J., Watson, R.T.: Analyzing the past to prepare for the future: writing a literature review. MIS Q. **26**(2), xiii–xxiii (2002)
69. Cooper, H., Hedges, L.: Research synthesis as a scientific process. In: Cooper, H., Hedges, L.V., Valentine, J.C. (eds.) The Handbook of Research Synthesis and Meta-Analysis, vol. 1. Russell Sage Foundation, New York City (2009)
70. Vom Brocke, J., Simons, A., Niehaves, B., Riemer, K., Plattfaut, R., Cleven, A.: Reconstructing the giant: on the importance of rigour in documenting the literature search process. In: 17th European Conference on Information Systems, pp. 2206–2217 (2009)
71. Rehurek, R., Sojka, P.: Software framework for topic modelling with large corpora. In: LREC Workshop on New Challenges for NLP Frameworks. ELRA (2010)
72. Feuerriegel, S., Fehrer, R.: Improving decision analytics with deep learning: the case of financial disclosures. In: 24th European Conference on Information Systems (2016)
73. Cerchiello, P., Nicola, G., Rönnqvist, S., Sarlin, P.: Deep learning bank distress from news and numerical financial data. In: DEM Working Paper Series (2017)
74. Rönnqvist, S., Sarlin, P.: Bank distress in the news: describing events through deep learning. Neurocomputing **264**, 57–70 (2017)
75. Diao, Q., Qiu, M., Wu, C.-Y., Smola, A.J., Jiang, J., Wang, C.: Jointly modeling aspects, ratings and sentiments for movie recommendation. In: 20th ACM SIGKDD International Conference on Knowledge Discovery and Data Mining, pp. 193–202. ACM (2014)
76. Yan, J., et al.: Mining social lending motivations for loan project recommendations. Expert Syst. Appl., 1–7 (2018)
77. Socher, R., Pennington, J., Huang, E.H., Ng, A.Y., Manning, C.D.: Semi-supervised recursive autoencoders for predicting sentiment distributions. In: Conference on Empirical Methods in Natural Language Processing, pp. 151–161 (2011)

Semantic Modelling

A Semantic Model Based Framework for Regulatory Reporting Process Management

Manjula Pilaka[(⊠)], Madhushi Bandara[(⊠)], and Eamon Mansoor[(⊠)]

School of Computer Science and Engineering, University of New South Wales,
Sydney 2052, Australia
manjulap@cse.unsw.edu.au, k.bandara@unsw.edu.au,
e.mansoor@student.unsw.edu.au

Abstract. As regulatory reporting involves loosely defined processes, it is a considerable challenge for data scientists and academics to extract instances of such processes from event records and analyse their characteristics e.g. whether they satisfy certain process compliance requirements. This paper proposes a software framework based on a semantic data model that helps in deriving and analysing regulatory reporting processes from event repositories. The key idea is in using business-like templates for expressing commonly used constraints associated with the definition of regulatory reporting processes and mapping these templates with those provided by an existing process definition language. A case study investigates the efficacy of the proposed solution in the case of an "Off-market bid" regulatory process. The results demonstrate the capability of the architecture in deriving process instances from a repository of Australian Company Announcements provided by the Australian Securities Exchange.

Keywords: Regulatory reporting · Process extraction · Semantic technology · Events

1 Introduction

Regulatory reporting processes help regulators in terms of monitoring the regulatory capital, safety and soundness of the legal entity, for example, ASX (Australian Stock Exchange) and ASIC (Australian Securities and Investments Commission) monitor the regulatory capital, safety and soundness of all the listed companies on ASX. They monitor risk associated with these legal entities, operations, and maintenance with the help of processes, procedures and policies specified by the regulators. They also monitor the cross-border transactions where there is involvement of multiple countries and ensure involved legal entities abide by the processes. e.g.: acquisition of an overseas company by an Australian company or vice versa. These regulatory reporting processes can impact the legal entities in many ways. For an example, failure to comply with required regulatory reporting processes can lead to civil penalties, legal and operational risks [1].

There are hundreds of data sources that can be used to collect different events that are pertaining to regulatory reporting such as company announcements, stock market feeds, news data. This paper is motivated by the need to better support analysts,

© Springer Nature Switzerland AG 2019
N. Mehandjiev and B. Saadouni (Eds.): FinanceCom 2018, LNBIP 345, pp. 149–164, 2019.
https://doi.org/10.1007/978-3-030-19037-8_10

academic researchers or data scientists in the finance domain who are interested in discovering important relationships between different events around specific regulatory reporting processes [2, 3]. Although there are many providers that offer different interfaces for analysing regulatory reporting event streams, it is still a considerable challenge to extract instances of regulatory reporting processes and analyse certain aspects like conformance of the process instance to the process model, variation of event instances within a process instances and assess the impact of process instances through third-party systems and additional data. Existing approaches do not deal adequately with the challenges as they demand both technical knowledge and domain expertise from the users. In addition, the level of abstraction provided does not extend to the concepts required by a typical data scientist or a business analyst.

The structure of the paper is as follows. Section 2 provides the literature review related to analysing the regulatory reporting processes with an example scenario of a regulatory reporting process including the expected output to be derived. Section 3 presents a semantic data model that can be used for deriving the regulatory reporting process and the proposed system architecture. Section 4 discusses validation of the architecture with a case study evaluation. Finally, Sect. 5 concludes with future research in this area.

2 Background and Related Work

The main difficulty in analysing regulatory reporting processes is that they are loosely framed processes i.e. there is no explicitly defined process model [4, 5]. Considering a process as a set of expectations, the problem is deriving a series of events that meet these expectations. For example, consider the "Off-market bid" process type shown in Fig. 1 which is a part of Mergers and Takeovers in Australia [6] and categorised as 'Takeover announcement'. Some of the event types that are part of the process are: announcing the bid, preparing and submitting bidder's statement to ASIC, submitting bidders statement to target stakeholders, submitting target statement, submitting accept or close offer announcement and compulsory acquisition announcement. An uncontested off-market bid process usually takes about 3 months, but in reality, it might take much longer time than anticipated. In cases where there is a counter bid, the timeline can be much longer. Even though one process instance might take longer time, it should be completed within 48 days of initiation in case of no extension or within 12 months of initiation in case of extension of the bid.

Each of the process types comprises of multiple event types and each event can impact on the performance of the company. As an example, Table 1 summarizes a process instance related to the off-market bid takeover of Genesis Resources LTD (GES) by Blumont Group LTD, a Singapore investment holding company in 2014 including the corresponding event instances and event flow along with the stock price and volume impacts. The impact is the difference in price or volume for each of the event instances from the previous event instance. As this acquisition has happened with extension of the bid it indicates the event flow in case of an extension of the bid. Table 2 reflects a comparison of expected and observed timelines for this regulatory reporting process instance.

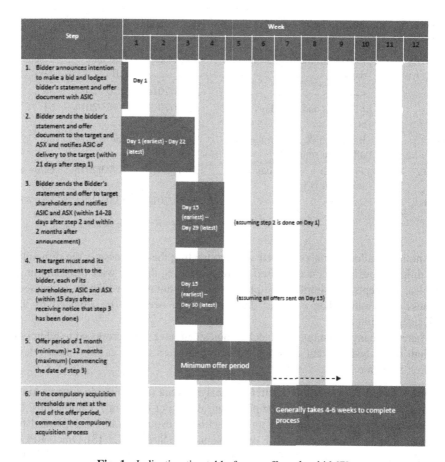

Fig. 1. Indicative timetable for an off-market bid [7]

Table 1. Instances of an off-market bid takeover process of GES by Blumont Group

Event type ID	Event type	Event instance date	Stock price impact (GES)	Stock volume impact (GES)
1	1001_Announcebid	24/01/2014		
2	1002_bidder's statement to ASIC	21/03/2014	−0.26	1232700
3	1002-Bidder's statement to target stakeholders	26/03/2014	−0.74	−447200
4	1003_Target statement	23/04/2014	−0.070002	−883700
5	1008_Offer extension	11/09/2014	−4.379999	157500
5	1008_Offer extension	29/10/2014	−0.67	−150000
5	1008_Offer extension	05/02/2015	−3.17	372500
5	1008_Offer extension	06/03/2015	−1.379999	496800
10	1009_Bid Unconditional	13/03/2015	0.289999	228700

Table 2. Compliance between expected and observed timelines for a regulatory reporting process instance

Between event types	Expected timeline (Business days)	Start date ABY	End date ABY	Calendar days
1–2	<21 days	21/01/2014	21/03/2014	58
1–3	>15 days and <29 days after 2 is sent	21/01/2014	26/03/2014	63
1–4	Before 15 days from the day offers are sent	21/01/2014	23/04/2014	88
4–10	>1 month from the day target statement is sent	23/04/2014	13/03/2015	Close to 1 yr
1–10	<1 year	21/01/2014	13/03/2015	1 yr 47 days

Another challenge analysts faced while extracting these process is the difficulty to map different business concepts used by rules and regulations with programming or technical concepts. A simple example is the time duration, which is expressed in business days rather than calendar days and hence needs to be adjusted for each analysis accordingly.

There are many techniques and technologies that have been proposed to address the problem of analysing event logs and extracting process instances. The most relevant research areas are:

Complex Event Processing (CEP): CEP analyses data from multiple sources to infer patterns of events that represent complex (causal, data or temporal) relationships. CEP provides event processing logic as an abstraction of event operations, separated from application logic. Some popular CEP platforms include Stream SQL [8], Oracle EPL [9], IBM Web sphere (business event processing system) [10] and Sybase Aleri CEP [11]. The major challenge in CEP is that their use involves a highly technical and time-consuming process. In addition, they lack the support of business concepts like business day derivation and business duration computation etc. An analyst would need have a good technical knowledge and expertise to build these concepts into the existing systems.

Process Mining & Conformance Checking: This is a relatively new research area [5, 12] and focuses on exploring, observing, and improving the overall business process based on analysing event logs that record activities performed by people, software, and machines. Process mining definition comprises of process discovery, conformance checking and model enhancement [13]. Process discovery is the discovery of processes from an event stream with the help of algorithms like a-priori, alpha and heuristic miner. Conformance checking means monitoring the processes periodically by comparing process model with process instances for any deviations or violations in processes. Model enhancement can be model checking and proposing new model or automated construction of simulation models. Existing process definition and mining tools in the market include Disco [14], Perceptive [15] and Celonis [16]. ProM [17] is the largest platform available for process mining and cater for different goals such as analysing social networks, exploring processes and validating business rules. The challenges with process mining tools are that they are not tailored to complex regulatory processes that would require inferred conditions (inferred condition means next event is determined by the presence or absence of an event), complex computations like conditions or inferred events and complex processing with business day computation concepts.

Process Definition and Modelling Languages: Since the nature of reporting processes renders formal notations like BPMN unsuitable, there are more appropriate processes modelling systems for loosely framed processes such as Flower (a Pallas Athen's case handling system) [18] and Tibco inconcert (an ad-hoc workflow system). In a case handling system, a process is a recipe for handling cases of any given type together with activities which are logical units of work to be executed when handling a case. Tibco Inconcert is a workflow system designed for unstructured processes. The systems specialty lies in creating new cases or process models from existing cases using templates that can be adapted to a single case or model a process while executing a case [19]. Another famous process modelling language is XML based Declare ConDec [20] which allows process extraction rules to be defined with the help of constraint templates. Each template has a graphical representation with a Linear Temporal Logic (LTL) formula specifying the semantics [21]. Declare ConDec is a constraint based Process Mining Relational language that allows for definition, verification, execution of constrained based process models and ad-hoc change of process instances. Major advantage of Declare is flexibility to design, change, specification of process models. These approaches are closely related to our work but limited in providing only technical abstractions for the analysts to express the rules related to process instance extraction.

In conclusion, existing approaches do not deal adequately with the challenges as they demand both technical knowledge and domain expertise from the users. We would still want to use and leverage existing technologies or systems as they are powerful, but we have observed that the existing solutions only partially address the requirements. In addition extensibility of the solution is required for new features like adding new business rules or integrating with third party systems An ideal solution would need to use both capabilities of CEP and process analysis technologies under the same framework in a complementary way.

3 Proposed Framework for Regulatory Reporting Process Management System (RPMS)

To solve the challenges and limitations discussed in Sect. 2, we propose a two-step solution: designing a semantic data model or an ontology representing events to define loosely framed processes and providing a system called RPMS (Regulatory Reporting Process Management System) that interprets the proposed semantic data meta-model to extract processes from event repositories. RPMS will be capable of defining a process and detecting corresponding process instances of regulatory reporting. In addition, it will also analyse the impact of the processes defined in semantic data meta-model.

This solution is focused towards the end-user programming or end-user software engineering [19]. End-user programming or software engineering can be defined as programming to achieve a result primarily for personal/research rather than public use with minimal programming experience or expertise from the experts and can provide the expert or domain knowledge in the simplest form. End-user software engineering (EUSE) focuses on approaches, techniques, and tools to improve the quality of end-user software [19].

The proposed semantic data model provides flexibility to represent events, dates in business days and define rules in the form of constraints or conditions. Semantic interoperability [24] enables not just the packaging of syntactical data but also the simultaneous transmission of unambiguous meaning of data through metadata. The linking of all the metadata is performed through a shared vocabulary.

3.1   Semantic Data Meta Model

To define the RPMS semantic data model we studied and identified a suitable set of ontologies that can represent different event instances, processes, and financial domain knowledge. To define events we chose the Adage-event model [22]. In addition, the Financial Industry Business Ontology (FIBO) [23] is selected to derive domain concepts such as business and financial dates. Finally we define a process model based on the Declare meta-models [20] to represent processes and templates. This section will discuss those models in more detail.

Adage – Event Meta Model

Although there are multiple data modelling frameworks available, there is no single standard for the representation of event data and event pattern occurrences. Every data modelling framework focuses on data from one particular domain or source and hence we have used Adage event meta modelling framework [22] dedicated to event data analysis, addressing the lack of standard in representation formats for event data and event patterns and lifting the barrier for data exchange between components. Adage event data modelling framework provides event data meta-model with a set of operational guidelines to help generate custom event data models, irrespective of sources or related domains. They propose Adage Financial Market data model which is flexible and generic enough to include datasets from sources like Bloomberg (2016), Thomson Reuters (2016), WRDS (2017). Listed below are the entities in ADAGE Financial Market conceptual data model relevant to our research.

- *Event*: Timestamped super class event which could represent any kind of Financial Event.
- *Market Event*: A subclass of event with a company that could represent as a superclass of any event across domains such as trade, quote etc.
- *Exchange*: Exchange is a platform for all traded products. Companies list and trade their products on the exchange. Exchanges maintain market datasets in a low-frequency form as in the end of day transaction datasets or high frequency form on time intervals less than a day on minutes or seconds basis [27].
- *Product*: There are two types of products- tradable (e.g. trade, quote, end of day events) or non-tradable (e.g.: index, news and market measure events) and they are distinguished by their product id.

- *Quote Event*: An exchange is responsible for generating Trade or Market Depth events. The quote is a time-stamped event which lists the buy and sell bids.
- *End of day event*: End of day events are time stamped events which represent values of trades on a daily basis.
- *Market Measure*: These are events that store time-stamped data related to different calculable measures such as daily returns and intraday returns.
- *Index*: Time-stamped events that represent a composite value of tradable products of a group or related to companies listed on the stock exchange. Some examples are the German DAX index listed on Frankfurt stock exchange and all ordinaries listed on the ASX [26].

For this research, we extended the above Adage Financial Market to include Australian Corporate Announcements that we refer to as ASX announcement events. Every ASX announcement event includes a company code, date, time, headline, and category code of the event which distinguishes the type of the announcement and the price sensitivity of the event (whether the announcement has impacted the stock price of the company on that date in a stock exchange). Therefore, an ASX Announcement class is created as a subclass of Market Event.

The major challenge with the current Adage financial market event is that it has a fixed timestamp which is not enough or adequate for concepts involving business dates, financial dates etc.

FIBO Meta Model

FIBO [23] is a recent initiative to define financial industry terms, definitions, and synonyms using semantic web principles. The initiative uses OMG Modelling standards which leads to transparency in global financial system and aid in regulatory reporting. It provides a regulation ontology, business readability to reports, and standardise the language to precisely define the terms, conditions, and characteristics of financial instruments, legal, business entities relationships, legal actions, and corporate process aspects in corporate actions. FIBO [23] addresses the regulatory transparency by defining standardised terminology ensuring interoperability among messaging protocols. It provides linked open data which ensures metadata annotations, regulatory rule conversion, linked data classes, class relationship data, structural, semantic and operation rules for interpretation of financial standards and regulatory requirements. This is done to enhance the monitoring of emerging institutional risks and systematic financial risks. The FIBO conceptual ontology is at a level of abstraction that intends to be a canonical model of primary business terms (concepts).

This paper focuses on the FIBO's Business Dates Module. This meta-model includes subject matter ontologies and owl classes for date and time concepts that are specific to financial services. As we would need the duration in Business day terms for regulatory compliance, this paper uses Business Day Ontology classes and each of its entities that are explained below:

- *Date*: A date identifies the calendar day.
- *Date Period*: A date span over one or more calendar days. This is defined by a minimum of two out of three of a start date, end date and duration.

- *BusinessDayConvention*: The number of possible ways in which a date can fall on a weekend or holiday can be handled. *BusinessDayTreatment* combines any date with *BusinessCenter* to determine the course of action when a business is conducted on a specific business day in a business centre.
- *BusinessDayAdjustment*: A BusinessDayAdjustment uses a *BusinessDayConvention* to specify what happens when a day falls on a day that is on a weekend or holiday in some business centre.
- *BusinessCentre*: the location where business is conducted and business calendar days are associated.

Process Model

To represent regulatory reporting processes in a flexible way, we adopt the approach used in loosely framed process languages which are based on the principle of defining processes as a set of constraints involving events, event entity conditions. Our process model will support a number of templates specifically adapted to regulatory reporting.

The proposed concepts are:

- *RPMSProcessType*: uniquely defines a regulatory process with its constraints and associated templates and parameters. For example: Offmarketbid is one type of regulatory reporting process and the process is identified with the help of constraints associated with this process.
- *RPMSConstraint*: represents the application of a RPMSTemplate with its parameters
- *RPMSTemplate*: represents a common set of parameters involving events and/or domain concepts. Such RPMS templates are implemented on top of Declare ConDec [20] templates. In other words, our templates can be considered as an application of Declare templates customised for the Regulatory Process Modelling domain.
- *RPMSParameters*: provide links to the actual parameters of the template which will be instances events and/or domain concepts. The list of RPMS templates and their expected parameters is shown in Table 3.

Table 3. Four types of RMPS templates and respective parameters

RPMS template	RPMS parameters	Role and function
Precedence	Event1, Event2, Duration	Determine the business day duration between two events
Alternate Precedence	Event1, Event2, Duration	Determine the alternate business day duration between two events
Absence	Event1	Determines if Event1 presence or absence in the event stream
Choice of 2 or 3	Event1, Event2, Event3, Duration1, Duration2	Determines if the next event is Event2 or Event3 and business day duration (either Duration1 or Duration2) between first event and next event based on presence or absence of Event1

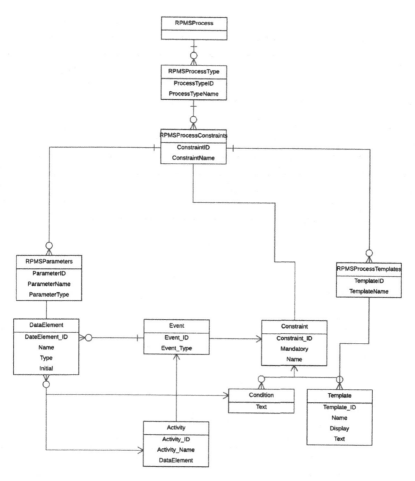

Fig. 2. Declare process model with constraints, activities and templates

Table 4. Mapping of RPMS templates with declare template

RPMS template	Declare template	Declare template parameters
Precedence	Precedence	Branches (Activity1, Activity2), condition (Duration), LTL Logic, Data Elements
Alternate Precedence	Alternate Precedence	Branches (Activity1, Activity2), condition (Duration), LTL Logic, Data Elements
Absence	Absence	LTL logic
Choice or 2 or 3	Choice or 2 or 3	Branches (Activity1, Activity2, Activity3), condition (Duration), LTL Logic, Data Elements

Declare templates expect different concepts as parameters. The relationships between the different concepts in the Declare language are represented in Fig. 2. In the proposed architecture, the RPMS templates are mapped to Declare Templates for execution by transforming RPMS parameters into the concepts expected by the Declare templates as parameters. The mapping between different RPMS process concepts and Declare template concepts are illustrated in the Table 4.

Proposed Ontology – Semantic Data Model

Figure 3 shows the integrated RPMS semantic data model that includes Adage Event model, FIBO Business Dates model and process model. We can see that the process modelling constraints or conditions can refer to business concepts. This includes ASX announcements, End of Day etc. as well as FIBO classes like business day convention which help derive the constraint verification based on business day conventions. This model also helps querying instance data in a linked data format.

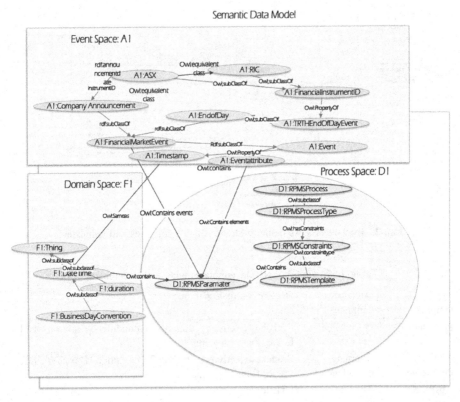

Fig. 3. Semantic meta data model

3.2   Architecture and Implementation

This section details the Regulatory Reporting Process Management System (RPMS) architecture, the second step of our proposed solution. We designed a five-layered architecture as shown in Fig. 4 and developed a prototype web application.

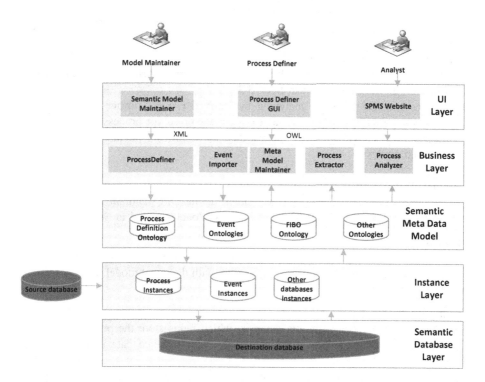

Fig. 4. Architecture for regulatory reporting process management system

UI (User Interface) Layer

The UI Layer has Process Definer Graphical User Interface (GUI) which helps in defining the process, the Semantic Model Maintainer GUI creates or maintains the semantic event model. In addition, the RPMS web application GUI is used by the analyst for Importing events, extracting processes, and analyzing the process with the assistance of process definer and semantic model maintainer.

Business Logic Layer

The Business Logic Layer includes several modules which are:

- *Process Definer module* helps store the process defined with the help of Process Definer GUI to the database. It helps to save the XML files obtained after defining a process with constraints using Declare designer [25] into the semantic database for further processing.

- *Event Importer* converts the event data file to RDF format or triples and helps store the data in the Semantic database.
- *Meta Model Maintainer* module is responsible for creating new event models and accessing or maintaining event model in the semantic database. The defined event model acts as a definition for the event instance data to be exported into the semantic database in the form of triples
- *Process extractor* module helps to extract process instances from the event instances (imported through the event importer module and stored in the semantic database) with the help of RPMS process templates. This is the core feature of the RPMS and involves a complex business logic based on *ProcessInstanceExtractor* algorithm discussed in detail in Sect. 3.3.
- *Process Analyzer* helps in analyzing the derived process further and visualizing results via the RPMS website. As an example, the prototype process analyzer can be integrate the behavior of stock market data for significant fluctuations during any identified process.

Semantic Data Meta Model Layer

The Semantic Data Meta Model Layer helps define and maintain the meta-model (ADAGE event meta-models, FIBO and Process Model) defined in Sect. 3.

Instance Layer

The Instance Layer helps to extract and convert the process, event data, and other data instances coming from different sources, to align with the event model in order to store within a database.

Database Layer

The Database Layer is a semantic database which helps store the process definition language, Semantic meta model, process, and event instance data required for the RPMS.

The implementation of the database layer was done using a MarkLogic semantic triple store. Through the Event Importer, instance data was loaded in RDF/TTL format, extracted and converted from various data sources like Sirca's Australian Corporate Announcements (ACA) and Thomson Reuters Tick History (TRTH) databases. This triple store also contains ontologies created by meta-model maintainer and other standard ontologies such as FIBO (Financial Industry Business Ontology) that help to describe the business concepts related to regulatory reporting.

3.3 Process Instance Extractor Algorithm

The process instance extractor algorithm takes a constraint (C) related to the process type to be extracted as an input. First, it loads the RPMS process template that is associated with the constraint. Secondly, the parameters associated with the constraint are extracted from the corresponding event instance data, incorporating FIBO and Adage event model instances.

For example, a precedence constraint which links to an RPMS Precedence template will have corresponding links to three parameters to be retrieved from the instance ontology (see Table 3): the Initial and End events from the Event ontology and the

Duration from the FIBO ontology. Then the parameters will be fed to the template before invoking the code associated with the RPMS Precedence template.

When this code is executed, a condition will be created (e.g. End Event Timestamp – Initial Event Timestamp < Duration), and two output activities are defined as outcomes based on whether the condition satisfy or not (e.g.- Events violate or satisfy the given constraint). The condition and outcomes defined would be passed the corresponding Declare template, which will execute the corresponding LTL Logic and the result will be stored in a tabular format for further analysis.

The main research contributions for this framework are

1. Framework provides an architecture for enabling analysts or academic researchers to extract and analyse instances of regulatory reporting processes specified at a high level of abstraction with a solution that leverages existing business knowledge which is stored in ontologies.
2. The framework helps analysts to define regulatory reporting process types with minimal technical knowledge and expertise leveraging existing technologies or ontologies available. Using Process Definer GUI and Semantic Meta Modeller the framework helps define the process type and semantic meta model with minimal technical knowledge and leveraging existing business knowledge and ontologies.
3. The framework is flexible enough to add new features, be interoperable with other existing systems and other services to do further analysis. With the Impact analysis the framework is flexible enough to be integrated with any other existing systems or databases like TRTH (Thomson Reuters) and do further analysis based on the needs of data scientists.

4   Case Study: Off-Market Bid Regulatory Process

This section showcases the capability of the proposed solution with a case study related to the Off-market bid process example provided in Sect. 2. For this process, we identified the relevant constraints or conditions through the Indicative Timetable of Off-market bid process (Fig. 1). Using those constraints and conditions the relevant rules were framed to design the off-market bid process through Declare's Process Definer GUI and the underlying semantic data model. The defined process is illustrated in Fig. 5.

For extracting the off-market bid process the relevant ASX announcement event instances were all Takeover announcements with Category Codes – 1001,1002, 1003,1008 and 1009 in the ACA Database. For the purpose of the case study, we extracted a data set of 24,000 announcements for 1495 companies over period from January-1995 to September. Event instances were converted to RDF triples format with the help of Event Importer module and added to the semantic database. The Process Extraction Module was then used to retrieve process instance data with the help of rule or constraints defined through the Process Definer GUI. Finally, the Process Analyzer module allows further analysis to be conducted using market data to identify the impact of the process instances of 'Off-market bid process' for related company and results can be shown regarding the stock price impact and stock volume impact occurred by any

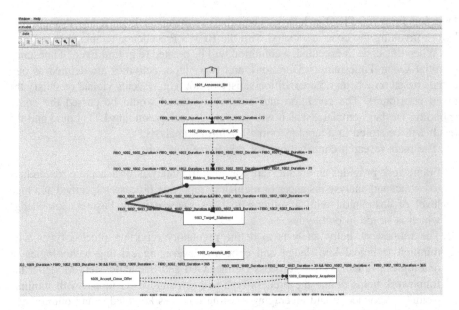

Fig. 5. Off-Market bid process constraints defined with declare designer user interface

particular process instance. As the market data is also imported though Event Importer adhering to same semantic meta-model, identifying process instances and conducting impact analysis is a single and integrated process for the user. Through the visualization facility, analysts can identify behaviour such as volume/price fluctuations and trading patterns within the identified regulatory process occurrence.

5   Conclusion and Future Work

This paper looks into the analysis of event data pertaining to regulatory reporting processes, discusses the challenges faced by the data analysts or researchers in deriving different process instances related to any event based system and outlines the limitations of existing approaches. Using a number of concepts related to complex event processing systems and process mining systems, it proposes a semantic process management system which helps define the processes related to any event model with a pre-defined set of constraints or conditions. With the aid of such process definitions and meta models, process instances can be extracted from a semantic database on-the-fly or on-demand. Due to the flexibility and linkability provided by semantic models, further analysis can be conducted incorporating identified processes and multitude of different data sets such as conducting impact analysis of the process using stock market data.

The proposed architecture can be extended to other domains incorporating different types of processes and event-based data by defining appropriate event subclasses and linking to external ontologies that define concepts that are relevant for the domain.

References

1. Stewart, J.: American banker - price of increased regulatory burden: less time for customers. https://www.americanbanker.com/news/price-of-increased-regulatory-burden-less-time-for-customers

2. Harrington, W., Heinzerling, L., Morgenstern, R.D. (eds.): Reforming Regulatory Impact Analysis. Routledge, Abingdon (2010)

3. Carroll, P.: Regulatory impact analysis: promise and reality. In: The ECPR/CRI Conference 'Frontiers of Regulation. Assessing Scholarly Debates and Policy Challenges'. University of Bath (2006)

4. Dumas, M., Van der Aalst, W.M., Ter Hofstede, A.H.: Process-Aware Information Systems: Bridging People and Software Through Process Technology. Wiley, New York (2005)

5. van der Aalst, W.M.P., La Rosa, M., Santoro, F.M.: Special issue on BPM "Use Cases". Bus. Inf. Syst. Eng. **58**(1), 1–6 (2016)

6. Riyanto, A.: The regulation of takeovers in Australia. Law Rev. **5**(3) (2013)

7. Introduction to Australia Takeovers. http://forms.minterellison.com/files/Uploads/Documents/Publications/Reports%20Guides/RG_2014_Intro_Aust_Takeovers.pdf

8. SQL Stream. http://sqlstream.com

9. Zang, C., Fan, Y.: Complex event processing in enterprise information systems based on RFID. Enterp. Inf. Syst. **1**(1), 3–23 (2007)

10. Maréchaux, J.L.: Combining service-oriented architecture and event-driven architecture using an enterprise service bus. In: IBM Developer Works, pp. 1269–1275 (2006)

11. Etzion, O., Magid, Y., Rabinovich, E., Skarbovsky, I., Zolotorevsky, N.: Context aware computing and its utilization in event-based systems. In: Proceedings of the Fourth ACM International Conference on Distributed Event-Based Systems. ACM (2010)

12. van der Aalst, W., et al.: Process mining manifesto. In: Daniel, F., Barkaoui, K., Dustdar, S. (eds.) BPM 2011. LNBIP, vol. 99, pp. 169–194. Springer, Heidelberg (2012). https://doi.org/10.1007/978-3-642-28108-2_19

13. van Der Aalst, W.: Process Mining: Discovery, Conformance and Enhancement of Business Processes. Springer, Heidelberg (2011). https://doi.org/10.1007/978-3-642-19345-3

14. Schmitt, B.: Process mining with Process Observer and Fluxicon Disco. SAP Community Network (2014)

15. van der Aalst, W.M.: Process Mining: Data Science in Action. Springer, Heidelberg (2016). https://doi.org/10.1007/978-3-662-49851-4

16. van der Aalst, W.M., Alfredo, B., van Zelst, S.J.: RapidProM: mine your processes and not just your data (2017)

17. van der Aalst, W.M.: Exploring the CSCW spectrum using process mining. Adv. Eng. Inform. **21**(2), 191–199 (2007)

18. van der Aalst, W.M., Weske, M., Grünbauer, D.: Case handling: a new paradigm for business process support. Data Knowl. Eng. **53**(2), 129–162 (2005)

19. Tzeremes, V., Gomaa, H.: XANA: an end user software product line framework for smart spaces. In: 49th Hawaii International Conference on System Sciences (HICSS). IEEE (2016)

20. Pesic, M., Schonenberg, H., Van der Aalst, M.P.: Declare: full support for loosely-structured processes. In: 11th IEEE International Enterprise Distributed Object Computing Conference, EDOC 2007. IEEE (2007)

21. Pesic, M., Schonenberg, M.H., Sidorova, N., van der Aalst, W.M.P.: Constraint-based workflow models: change made easy. In: Meersman, R., Tari, Z. (eds.) OTM 2007. LNCS, vol. 4803, pp. 77–94. Springer, Heidelberg (2007). https://doi.org/10.1007/978-3-540-76848-7_7

22. Weisi, C., Rabhi, F.A.: Enabling user-driven rule management in event data analysis. Inf. Syst. Front. **18**(3), 511–528 (2016)
23. Newman, D.: The Financial Industry Business Ontology (2013)
24. Hutchison, D., Mitchell, J.C.: The Semantic Web (1973)
25. Westergaard, M., Maggi, F.M.: Declare: a tool suite for declarative workflow modelling and enactment. BPM (Demos) **820**, 1–5 (2011)
26. Rabhi, F.A., Guabtni, A., Yao, L.: A data model for processing financial market and news data. Int. J. Electron. Financ. **3**(4), 387–403 (2009)
27. Tsay, R.S.: Analysis of Financial Time Series, vol. 543. Wiley, New York (2005)

Applying Ontology-Informed Lattice Reduction Using the Discrimination Power Index to Financial Domain

Qudamah Quboa[1]([⊠]), Nikolay Mehandjiev[1], and Ali Behnaz[2]

[1] Alliance Manchester Business School, University of Manchester,
Manchester, UK
{qudamah.quboa, n.mehandjiev}@manchester.ac.uk
[2] School of Computer Science and Engineering,
University of New South Wales, Sydney, Australia
ali.behnaz@unsw.edu.au

Abstract. Contemporary financial institutions are relying on varied and voluminous data and so they need advanced technologies to provide their customers with the best possible services. Capturing the meaning, or semantics, of data and presenting these semantics in simplified yet relevant models are key challenges to achieving this. Formal Concept Analysis (FCA) automates the analysis of properties and instances of the data, generating a lattice which groups properties and instances into concepts. This lattice can be used as automatically generated semantic structure describing the domain, yet the complexity and size of the resultant lattice render this technique unusable in most practical cases involving financial data. To tackle this, our Ontology-informed Lattice Reduction approach can guide the reduction of the lattices generated from financial sampled data. We validate the adaptation of the approach to the financial domain through a real-world asset allocation case study, demonstrating that the approach achieves good overall performance and relevant results.

Keywords: FCA · Semantic structures · Lattice reduction · Validation

1 Introduction

The financial industry has been undergoing immense changes and disruption in the last decade. Indeed, one can argue that no force has influenced the industry more than abundant data and cheap computational power. Modern financial institutions are focused on excellence through advanced information technologies. To provide state-of-the-art customer service, financial institutions collect vast amount of data including news data, financial market data, sentiment data and data from other exogenous factors. However, the heterogeneous nature of this data makes it difficult for players in financial industry to obtain real benefits from their data. One key obstacle in leveraging data assets is difficulties in capturing the meaning (semantics) of the data. Therefore, knowledge acquisition is a key step in the process and its automation will be an enabler. Automating knowledge acquisition involves [1], processing voluminous data

© Springer Nature Switzerland AG 2019
N. Mehandjiev and B. Saadouni (Eds.): FinanceCom 2018, LNBIP 345, pp. 165–179, 2019.
https://doi.org/10.1007/978-3-030-19037-8_11

[2], understanding the meaning of this data and its relations [3] and presenting the outcomes in relevant yet simplified models [4].

For practical use, manual semantic tagging of data is very expensive task and many scientific researchers are working to automate this task. As an example, most data in financial markets has high volume and needs to be processed at a high speed to provide value for its users. Formal Concept Analysis (FCA) is one technique for tagging semantics to data. FCA takes a matrix of incidence relationships between sampled data properties (intent) and their object instances (extent), named a formal context, and then builds a lattice of partial order relations between the two sets (instance and property sets). One major issue here is the complexity and nosiness of lattices produced by FCA to be used for practical semantic analysis of real-world datasets. To solve this issue, a reduction of lattices is required and existing approaches to achieve that are based on mathematical measurements of relevancy [4]. They are agnostic about any prior knowledge regarding the targeted domain, even when it is already formalised and represented in an ontology or semantic structure.

Inspired by the similarities between ontology-based semantic and FCA representations, different approaches have been proposed to use the combination of both such as ontology modelling and attribute exploration [5, 6] and merging different ontologies [7, 8]. However, the use of existing domain knowledge (represented as ontology) to support the lattice reduction has not been explored until now. In our research [9], we proposed Ontology-informed Lattice Reduction approach that address the attaching of semantics to further instances in the domain through FCA using existing domain ontology.

The approach uses prior domain knowledge (encoded in semantic format - ontology) to classify and guide the reduction process of a sampled formal context where not all instances are in the ontology. In addition, the approach relies on a new relevancy metric called Discrimination Power Index (DPI) that is used to automatically classify any new instances based on the shared instances and the overall power of a property within the formal context.

In this paper, we present a more detailed analysis and testing of the proposed approach [9] to evaluate and validate the performance of the approach especially when facing the problem of large lattices generated when applying FCA to real-world data. A financial real-world case study is presented to confirm the feasibility and the validity of the approach.

The remainder of this paper is structured as follows: Sect. 2 describes shortly the background of the work. Section 3 summarises briefly the proposed approach and its stages. Section 4 introduces a real-world case study in financial markets to evaluate the adaption of the approach. Then Sect. 5 discusses the results and the statistical analysis of applying the approach and Sect. 6 concludes the paper.

2 Literature Review

2.1 Analytics in Financial Domain

As data is becoming more abundant, organizations are looking for ways to acquire actionable insights from their data, and hence make better decisions, achieve value and competitiveness. Domain experts and analytics units at organizations have access to sophisticated analytics solutions which serve their users' ways of conducting analytics.

For instance, a bank would be interested to forecast price moves in a particular asset. There are thousands of assets (instances) to be analysed, and for each asset hundreds of properties (intents) can be input into a predictive model to forecast the changes in the respective asset [10, 11].

A typical analytics problem in finance domain has heterogeneous complicated datasets which need to be understood easily and acted upon. As a result, knowledge acquisition and knowledge representation are key elements in enhancing big data analytics problems. Semantic web technology and ontologies present a solution to capture knowledge in a domain [12].

2.2 Semantic Web Ontology

Ontology is a well-known knowledge representation method and widely supported by both the academic and the industry domains in terms of available software and tools. Ontologies defines as "explicit formal specifications of the terms in the domain and relations among them" [12]. Different general and specialist ontologies in various domains have developed by experts to capture knowledge and pass information in standardised way.

It is mainly designed to define a set of data and its structure, constrains for the use of other applications, and commonly utilised as a data sharing mechanism between various programs or software agents. For instance, FIBO (the Financial Industry Business Ontology) is an "industry initiative to define financial industry terms, definitions and synonyms using semantic web principles [13]."

2.3 Formal Concept Analysis (FCA)

Formal Concept Analysis (FCA) is a mathematical formalism to automatically analyse the structure of a domain of interest [2]. It generates a lattice of concepts representing incidence relations between sets of observed properties (intent) and their object instances (extent) in a target domain. The generated lattice is constituted by formal concepts produced from mapping these relationships onto a knowledge structure that reflects the specialisation and generalisation among the concepts of resultant lattice [2].

"Sports and their attributes" [14] (presented in Table 1) is an example of formal context (K). In this example, the extent set (G) is {Run, Gymnastics, Triathlon, Football, Tennis, Baseball, Curling, Diving, Rowing}, the intent set (M) is {on land, on ice, in water, collective sport, individual sport, using ball, needs opponent, multiple disciplines} and the sign X identifies a relation between instances of G and properties of M (representing the set of I).

Table 1. "Sports and their attributes" formal context (taken from [14]).

	On land	On ice	In water	Collective sport	Individual sport	Using ball	Needs opponent	Multiple disciplines
Run	X				X			
Gymnastics	X				X			X
Triathlon	X		X		X			X
Football	X			X		X	X	
Tennis	X			X		X	X	
Baseball	X			X		X	X	
Curling		X		X				
Diving			X		X			
Rowing			X	X				

By processing the relationships between properties and instances of this example, the corresponding concept lattice will be formed, as shown in Fig. 1, where each node represents a formal concept and each connecting edge represents a *subconcept-superconcept* relationship.

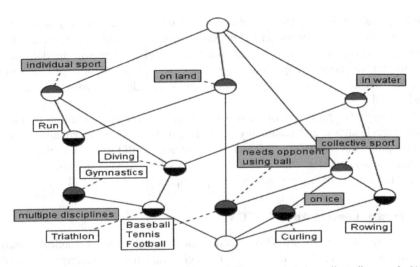

Fig. 1. Concept explorer-generated FCA lattice of "Sports and their attributes" example [14].

2.4 Existing Lattice Reduction Techniques

A bottleneck problem with FCA mechanism is the huge size of lattices created from real-world data sets because of noise and exceptions [2, 15]. Various methods have been proposed to reduce the lattice using the structure of the lattice itself. These are categorised into redundant information removal, simplification or selection [4].

The focus of the work is the last category, the selection reduction, which represents any approach that emphases on selecting specific properties, instances and/or concepts based on different measurements of relevancy (a set of constraints that requires to be satisfied). This could be depended on various measurements, such as logic (according to a user's attributes priorities [16]), weight (frequent weighted concept reduction [17]), or hierarchies (using hierarchically ordered attributes [18]). This kind of reduction is performed after completing the construction of the formal context [4].

All these approaches mainly rely on the lattice structure and are agnostic about any prior knowledge about the domain that makes the results more vulnerable to systemic noise in the data.

2.5   Similarity Measurements

Three similarity measurements are used to align the two different formal contexts (ontology-derived and sampled) and integrate them. The first two are:

Jaccard Similarity Coefficient Index. This well-known similarity measurement [19] is based on the following formula:

$$Jaccard\,Index\,S_{Jac} = \frac{|B_1 \cap B_2|}{|B_1 \cup B_2|} \tag{1}$$

Hamming Distance Index. This is also a well-known similarity measurement [20] and is based on the following formula:

$$D_{hamming} = b + c \tag{2}$$

It is worth to mention that these two measurements weight all properties as of equal importance whilst our observations reveal that specific properties could have a higher discrimination power than others.

We thus introduced a new complementary index called **Discrimination Power Index (DPI)** [9] to identify a unique most-similar concept. It enhances the similarity selection process in picking one of the possible concepts (pre-filtered from the previous indices) based on their properties' overall discrimination power. This is formally defined as [9]:

$$DPI = \frac{|\{\forall g \in G \,|\, b \in B_1 \cap B_2 : (g, b) \in I\}|}{|G|} \tag{3}$$

The theoretical explanations of each of the mentioned equations are covered in [9].

3   Ontology-Informed Lattice Reduction Approach

Our approach presented in [9] uses existing knowledge about the target domain (encoded in a semantic ontology format) to support reduce the FCA-generated lattice when extracting structure from sampled formal context (data). The approach begins by extracting and transforming all required information into acceptable formats. It then starts by recognising object instances that exist in the formal context and the ontology-derived context and aligns concepts using these identified instances. This is followed by automatically structures the rest of object instances from the formal context, using (a) basic alignment routine that rely on the properties they have in common with the shared instances and (b) advanced alignment routine that is based on the similarity measurements and the discrimination power of the properties. At the end, the resultant extended structure will be reduced based on the user's reduction threshold. The general outline of the approach shown in Fig. 2.

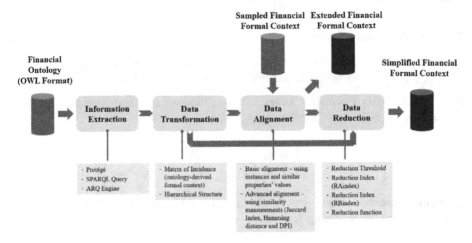

Fig. 2. The outline of the Ontology-informed Lattice Reduction approach [adapted from 9].

In the next subsections, a summary of each stage of the proposed approach (presented in Fig. 2) is provided.

3.1 Data Extraction

During this stage, Protégé, a well-known Ontology editor and knowledge acquisition software, is used to establish a deep understanding of the domain ontology as well as helping in the construction of the constrains of the retrieval queries.

To retrieve the required features and their instances from any Semantic Web dataset, it is necessary to use a semantic query language. SPARQL, recommended by W3C, is used as a simple protocol and Semantic Web query language to perform the querying via pattern matching [21].

Lastly, ARQ engine is a query engine for Jena (Java framework for building Semantic Web applications), which provides the support to the standard SPARQL query language, is used to execute the SPARQL queries, extract the results, and save them into temporary comma-separated values (csv) file that will be used in the next stage.

3.2 Data Transformation

The main purpose of the stage is to reformat the extracted information from the ontology source into a formal context format (in a tabulated format) and support it with transforming the hierarchal structure of the extracted data based on the ontology source itself and storing the results in a matrix format to easy the access.

The developed algorithms for this stage support multiple hierarchal levels of ontology structure and for any number of instances. This stage works as a prepossessing stage to complete all the required preparations to align the different datasets and reducing the results later on.

3.3   Data Alignment

Initially, a basic matching routine is applied based on instances existing in both the ontology and the sampled formal context. This is followed by a more complex classification routine for any instance not existing in the ontology and none of the classified instances shares the same properties. This is achieved by assigning any unknown concept from the sampled formal context to one of the existing ontology concepts using a combination of different similarity indices to evaluate and align concepts from both contexts based on their intents and extents.

This advanced alignment routine starts with a Jaccard similarity coefficient index followed by optional use of a Hamming distance, and then the proposed DPI [9] (again optional). The outcome of this stage is an extended formal context incorporating both the ontology-derived knowledge and the sampled formal context.

3.4   Data Reduction

Two indices are proposed in [9] to provide an indication whether a property is essential to a specific concept from the ontology or not:

RAindex is the first reduction index that focuses on evaluating the weight of every property in the sampled formal context regarding each and every extracted concepts from the domain ontology.

RBindex is a complementary index to indicate the concepts that need the reduction by relying on both the property and the ontology's concept in making the call.

Depending on the outcomes of both *RAindex* and *RBindex* indices, the reduction function is executed when required. The purpose of this function is to eliminate the incidence relations between the assessed property related to the used ontology concept and the object instances.

The theoretical explanations and the justifications of each of the mentioned indices and the reduction function are covered in [9]. This stage depends mainly on the sampled formal context and the extracted ontology information and hierarchical structure (prepared in the data extraction and data transformation stages) to work out which of the incidence relationships need to be removed from the extended formal context (resulted from the data alignment stage).

4   Financial Assets Allocation Case Study (Ontology and Formal Context of Exchange Traded Funds)

In this case study, we use semantic technology to capture and represent the knowledge related to asset allocation. Asset allocation intends to distribute investment among different financial assets so as to achieve a certain investment strategy [22]. The knowledge is extracted and represented in an ontology including a list of Exchange Traded Funds (ETF[1]). The goal in this case is to design and build an automated financial advice system which helps people invest and manage their funds at a fraction of the cost for human financial advice. As part of asset allocation, we explore, analyse and select a number of assets (in this case ETFs). We have obtained the data for these ETFs from Bloomberg. In doing so, we have adopted Bloomberg terminology for assets, their properties and respective categories. The sample consists of 87 ETFs (instances) and 42 (properties) presented in the Asset Selection ontology (see Fig. 3). Table 2 provides simple statistics for both the ontology and sampled formal context then Table 3 presents the data description of the financial sampled formal context properties.

Table 2. Basic statistical analysis of the financial sampled formal context and its ontology.

	No of instances	No of properties (Classes)	No of concepts	No of levels/height	No of edges
Sampled formal context	87	36	546	9	1722
Ontology	87	42	44	4	74

Table 3. The data description of the sampled financial formal context.

Properties	Possible values
Closed for new creations	{Yes, No}
Leverage/Short	{Yes, No}
Invests in Derivatives	{Yes, No}
Invests in swaps	{Yes, No}
Invests in physical commodities	{Yes, No}
Actively managed	{Yes, No}
Currency hedged	{Yes, No}
Index replication strategy	{Full, Optimized, Not Applicable, Derivative}
Index weighting methodology	{Market Cap, Not Applicable, Single Asset, Multi Factor, Fundamentals, Dividend, Proprietary, Equal}
Rebalancing frequency	{Quarterly, Not Applicable, Yearly, Semi-Annually, Other, Monthly}
Creation/Redemption	{In-kind, In-kind/Cash, Not Applicable, Cash}
Dividend frequency	{Quarter, Semi-Anl, Annual, None, Monthly, Irreg}
Risk	{Low, High}

---

[1] Exchange Traded Funds or ETFs are a basket of other assets that are designed to trace the performance of an index.

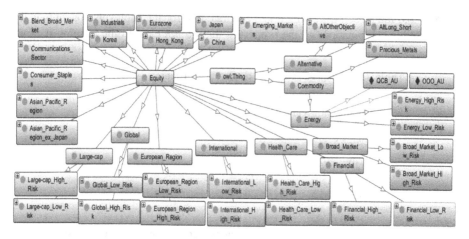

Fig. 3. Asset selection ontology.

5   Analysis and Discussion

To validate the validity of applying the approach in this financial context, different validation measurements are carried out to ensure each and every stage is generating the expected outcome with the right results.

By relying on Protégé, the construction of the SPARQL query is achieved and the retrieved results (using ARQ Engine) have all the main classes and sub-classes and their hierarchal structure levels within the domain ontology. The extracted information is then compared with the ontology itself to confirm the correctness of outcome. This includes matching the number of extracted instances (objects) and their classes-subclasses properties.

Moving to the second phase, Data Transformation, the validation here is achieved by comparing the outcome of the previous stage with the transformed results and making sure that all the data are transformed and correctly aligned.

During the Data alignment stage, two measurements are used to evaluate and validate the process: (1) testing the functionality of the basic alignment routine alone, and (2) testing the efficiency of alignment process with the advanced alignment routine (the Jaccard index, Hamming Distance, and DPI).

For the first test, the experiment starts by using only known instances that exist in both the financial sampled formal context and the financial ontology. Then the validation of the alignment routine is confirmed by comparing the outcome of this stage (Extended financial formal context) with the original sources and ensuring that all the instances are classified accurately as they should be.

For the second evaluation, 50% of the ontology instances are removed to create unknown instances in the sampled financial formal context while the other 50% are kept to construct the training part for the advanced alignment routine to work. Then the resulted classification outcome is evaluated and compared with the original one to confirm the performance of the alignment. Table 4 presents a summary of the results. It is worth to mention that the accuracy of the alignment for the top level of the ontology is 100% and the reason of having a less accurate rate for the sublevels concepts is the large possible combinations of the properties that do not exist in the training dataset.

Table 4. Alignment validation results (50% Training - 50% Testing).

Semantics attachment alignment accuracy	No of instances	Overall percentage
100%	54	62%
95%	1	1%
93%	10	11%
90%	22	25%
Below 90%	0	0%

During the last stage of the applying the approach, the performance of the reduction function is evaluated using two different extended formal contexts: (a) the extended one using the basic alignment routine and (b) the extended one using both alignments

Table 5. The reduction performance using the fully classified formal context (the extended formal context) based on the basic alignment routine only.

Reduction threshold	No of the concepts of the formal Context	Actual reduction %	Total no of edges	No of levels/height	Total no of concepts with the ontology's concepts	No of ontology concepts	No of overlapped concepts
90%	24	95.6%	40	5	59	44	9
80%	24	95.6%	40	5	59	44	9
70%	24	95.6%	40	4	61	44	7
60%	31	94.3%	54	5	85	44	10
50%	97	82.2%	248	8	203	44	62
40%	148	72.9%	414	8	293	44	101
30%	269	50.7%	849	9	487	44	174
20%	324	40.7%	1030	9	598	44	230
10%	477	12.6%	1519	9	801	44	280
0%	546	0.0%	1722	9	878	44	288

Table 6. The reduction performances using the extended formal context based on the basic and advanced alignments routines (50% Training - 50% Testing).

Reduction threshold	No of the concepts of the formal context	Actual reduction %	Total no of edges	No of levels/height	Total no of concepts with the ontology's concepts	No of ontology concepts	No of overlapped concepts
90%	24	95.6%	39	4	57	44	11
80%	24	95.6%	39	4	59	44	9
70%	30	94.5%	48	5	69	44	5
60%	43	92.1%	76	6	92	44	5
50%	121	77.8%	315	8	214	44	49
40%	185	66.1%	533	8	303	44	74
30%	313	42.7%	984	9	477	44	120
20%	334	38.8%	1058	9	520	44	142
10%	486	11.0%	1546	9	682	44	152
0%	546	0.0%	1722	9	790	44	200

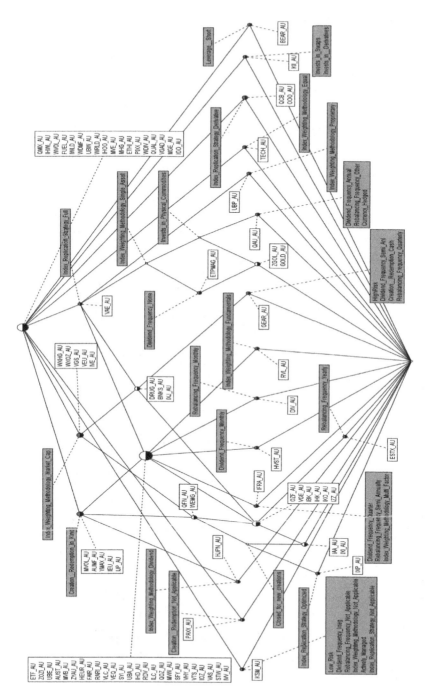

Fig. 4. The actual reduced sampled formal context at 60% reduction threshold (without the ontology attachments).

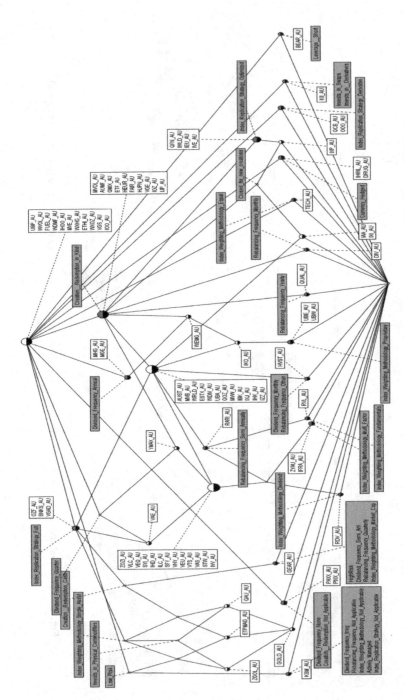

Fig. 5. The actual reduced sampled formal context at 60% reduction threshold (without the ontology attachments) based on 50% Training and 50% Testing.

routines (50% Training – 50% Testing). The performance of the function is evaluated at various reduction thresholds (presented in Tables 5 and 6 respectively).

Figures 4 and 5 illustrate the generated lattice of the reduced sampled formal context at a reduction threshold of 60% for both alignments experiments respectively without the semantic attachments.

It could be noticed that the performance of the approach using the advanced alignment routine (training and testing datasets) is very similar to the actual known one, which reconfirms the efficiency of the alignment stage.

In addition, the reduction function is working as expected and performing well in reducing the sampled data and presenting the results using simplified (and relevant) financial lattice, even when the knowledge base only provides partial coverage of the domain of interest.

It is worth to mention that (1) the reduction threshold represents the minimum ratio to pass the *RBindex* and not being flagged as an unnecessary concept. (2) The actual reduction is different from the reduction threshold and its value is vary depending on the incidence relationships of the actual formal context and its semantic extension.

6 Conclusion

The financial institutions gather vast amount of data from various resources including financial market data and news data. However, to gain an advantage as a player in the market and obtain real benefits, the understanding of the meaning (semantics) of the data and the presentations of outcomes in simplified and relevant models are the key obstacles that need to be solved. Formal Concept Analysis (FCA) helps with the semantics by generating a lattice that comprises partial order relations between sets of properties (intent) and their instances (extent) in a domain that maps onto a semantic structure. The problem is the resultant lattice is too complex and noisy.

Using existing domain knowledge to inform and reduce a formal context (that is taken by FCA) is an opportunity that is being utilised in this work to simplify the resultant lattice and presents relevant models.

In this research, the Ontology-informed Lattice Reduction approach is applied to the financial domain as this approach relies on the use of an existing ontology to inform and reduce the financial sampled formal context based on different alignments and reductions measurements.

We specifically apply the approach to asset allocation problem in financial markets and assess the feasibility and validity of the different stages of the approach and the performance of the approach in regards to this real-world case study. The achieved results are good and pass all the testing measurements in producing creating a simplified, yet relevant, result that could be used in practice.

In the future work, we will (1) Extend the reduction approach to include the reasoning of ontology's description logics (DLs) to increase the accuracy of the reduction constrains. (2) Continue the work on the semantic attachments and add a "loopback" feature as a new extension that permits the utilisation of the approach outcomes to enrich the existing ontology.

References

1. De Mauro, A., Greco, M., Grimaldi, M.: A formal definition of big data based on its essential features. Libr. Rev. **65**(3), 122–135 (2016)
2. Singh, P.K., Kumar, C.A., Gani, A.: A comprehensive survey on formal concept analysis, its research trends and applications. Int. J. Appl. Math. Comput. Sci. **26**(2), 495–516 (2016)
3. Rouane, M.H., Huchard, M., Napoli, A., Valtchev, P.: A proposal for combining formal concept analysis and description logics for mining relational data. In: Kuznetsov, S.O., Schmidt, S. (eds.) ICFCA 2007. LNCS (LNAI), vol. 4390, pp. 51–65. Springer, Heidelberg (2007). https://doi.org/10.1007/978-3-540-70901-5_4
4. Dias, S.M., Vieira, N.J.: Concept lattices reduction: definition, analysis and classification. Expert Syst. Appl. **42**(20), 7084–7097 (2015)
5. Ignatov, D.I.: Introduction to formal concept analysis and its applications in information retrieval and related fields. In: Braslavski, P., Karpov, N., Worring, M., Volkovich, Y., Ignatov, D.I. (eds.) RuSSIR 2014. CCIS, vol. 505, pp. 42–141. Springer, Cham (2015). https://doi.org/10.1007/978-3-319-25485-2_3
6. Baader, F., Ganter, B., Sertkaya, B., Sattler, U.: Completing description logic knowledge bases using formal concept analysis. In: Proceedings of the 20th International Joint Conference on Artificial Intelligence (IJCAI), Hyderabad, India, pp. 230–235 (2007)
7. Stumme, G.: Using ontologies and formal concept analysis for organizing business knowledge. In: Becker, J., Knackstedt, R. (eds.) Wissensmanagement mit Referenzmodellen, pp. 163–174. Physica, Heidelberg (2002)
8. Sarmah, A.K., Hazarika, S.M., Sinha, S.K.: Formal concept analysis: current trends and directions. Artif. Intell. Rev. **44**(1), 47–86 (2015)
9. Quboa, Q., Behnaz, A., Mehandjiev, N., Rabhi, F.: Ontology-informed lattice reduction using the discrimination power index. In: Proceedings of the 24th International Conference on Conceptual Structures (ICCS), Marburg, Germany, July 2019
10. Behnaz, A., Natarajan, A., Rabhi, F.A., Peat, M.: A semantic-based analytics architecture and its application to commodity pricing. In: Feuerriegel, S., Neumann, D. (eds.) FinanceCom 2016. LNBIP, vol. 276, pp. 17–31. Springer, Cham (2017). https://doi.org/10.1007/978-3-319-52764-2_2
11. LaValle, S., Lesser, E., Shockley, R., Hopkins, M.S., Kruschwitz, N.: Big data, analytics and the path from insights to value. MIT Sloan Manag. Rev. **52**(2), 21–32 (2011)
12. Gruber, T.R.: A translation approach to portable ontology specifications. Knowl. Acquis. **5**(2), 199–220 (1993)
13. Financial Services Standards. http://www.omg.org/hot-topics/finance.htm. Accessed 19 Apr 2018
14. Belohlavek, R., Trnecka, M.: Basic level of concepts in formal concept analysis. In: Domenach, F., Ignatov, D.I., Poelmans, J. (eds.) ICFCA 2012. LNCS (LNAI), vol. 7278, pp. 28–44. Springer, Heidelberg (2012). https://doi.org/10.1007/978-3-642-29892-9_9
15. Singh, P.K., Kumar, C.A.: Concept lattice reduction using different subset of attributes as information granules. Granul. Comput. **2**(3), 159–173 (2017)
16. Belohlavek, R., Vychodil, V.: Formal concept analysis with background knowledge: attribute priorities. IEEE Trans. Syst. Man Cybern. Part C Appl. Rev. **39**(4), 399–409 (2009)
17. Zhang, S., Guo, P., Zhang, J., Wang, X., Pedrycz, W.: A completeness analysis of frequent weighted concept lattices and their algebraic properties. Data Knowl. Eng. **81**, 104–117 (2012)
18. Bělohlávek, R., Sklenář, V., Zacpal, J.: Formal concept analysis with hierarchically ordered attributes. Int. J. Gen. Syst. **33**(4), 383–394 (2004)

19. Domenach, F., Portides, G.: Similarity measures on concept lattices. In: Wilhelm, A.F.X., Kestler, H.A. (eds.) Analysis of Large and Complex Data. SCDAKO, pp. 159–169. Springer, Cham (2016). https://doi.org/10.1007/978-3-319-25226-1_14
20. Choi, S.S., Cha, S.H., Tappert, C.C.: A survey of binary similarity and distance measures. J. Syst. Cybern. Inf. **8**(1), 43–48 (2010)
21. W3C, SPARQL 1.1 Query Language. https://www.w3.org/TR/sparql11-query. Accessed 23 Apr 2018
22. Sharpe, W.F.: Asset allocation: management style and performance measurement. J. Portfolio Manag. **18**(2), 7–19 (1992)

A Statistical Learning Ontology for Managing Analytics Knowledge

Ali Behnaz[1]([⊠]), Madhushi Bandara[1], Fethi A. Rabhi[1],
and Maurice Peat[2]

[1] School of Computer Science and Engineering,
University of New South Wales, Sydney, NSW 2052, Australia
ali.behnaz@student.unsw.edu.au,
{k.bandara, f.rabhi}@unsw.edu.au
[2] The University of Sydney Business School, Sydney, NSW 2006, Australia
maurice.peat@sydney.edu.au

Abstract. This paper focuses on the use of knowledge management techniques to help organisations tap into the power of statistical learning when conducting analytics. Its main contribution is in the use of an ontology development process to derive the essential concepts required for an ontology to represent variables of interest and their interrelationships with each other and with statistical datasets. This ontology is developed with the help of two case studies in the area of digital marketing and commodity pricing. A number of competency questions have been designed to map to user requirements in both case studies. A prototype system has been developed using a semantic modelling tool and a semantic data repository to demonstrate that the proposed ontology can support the competency questions via semantic queries.

Keywords: Statistical learning · Data science · Computational social scientist · Ontology · Semantic technology

1 Introduction

Data has long been utilized by researchers to run empirical tests or descriptive analysis. Increasing development in computer technologies in both software and hardware has enabled the storing, processing and exploring huge datasets (aka. Big Data) in the last decade. Analysis of these datasets opens up a new generation of big data applications that are of huge interest to researchers and companies in various areas. The increasing amount of data now covers different aspects of our lives. This data span from purchase habits, travels, friendship on social media, and beyond to collected data from sensors and machines. This availability of data has enabled us to predict human behaviour accurately, which seemed impossible even a decade ago. Predicting human behaviours and patterns is a central study in the area of social sciences. Through democratisation of analytics and data science, different domain experts are applying data science to their area to extract insights or enhance data-driven decision making. For instance, computational social scientists are experts who study behavioural and social dynamics via social network analysis, social media analysis and social simulation, all thanks to advances in analytics and data science [1, 2].

© Springer Nature Switzerland AG 2019
N. Mehandjiev and B. Saadouni (Eds.): FinanceCom 2018, LNBIP 345, pp. 180–194, 2019.
https://doi.org/10.1007/978-3-030-19037-8_12

Data science is increasingly being used alongside big data [3]. One definition states that data science entails systematic enterprise which organises and builds knowledge in the form of predictions and explanations. There are multiple user types who will be involved in data science production for enterprises. These user types will need to apply statistical learning techniques. A conventional data scientist needs to apply sophisticated machine learning and statistical learning skills to deliver actionable insights to a business. Gartner, Inc. believes that more than 40% of data science function will be carried out by machines by 2020, resulting in further democratisation of analytics by citizen data scientists. A citizen data scientist, as defined by Gartner, is an expert in creating or generating statistical learning models with the purpose of diagnostic analytics or predictive and prescriptive analytics [4].

There are multiple analytics frameworks and platforms for data management, data mining, or statistical analysis (e.g. SAS, SPSS, Matlab). Also, there exist hundreds of libraries in programming languages such as R or Python for statistical learning. However, the problem arises when analysts are trying to design a flexible analytics process which involves the use of a mix of frameworks, process steps and decision logic based on different abstractions. Most organisations rely on manual interventions to achieve integration between analytics tasks and their IT infrastructure via programs or scripts [5]. These types of solutions are tedious, expensive and hard to maintain.

In this paper, we focus on statistical learning as the backbone of most analytics problems. Then we analyse how organisations can benefit from statistical learning and why they need to build data science expertise. Next, we analyse existing solutions for data integration and knowledge capture. In the following section, we present our approach for dealing with data and knowledge disparity. We then introduce our ontology and in the following sections validate our research via competency questions and inference queries. Finally, we conclude by discussion and laying out future research.

2 Background and Related Work

To support statistical learning, organisations need well-established processes to acquire meaningful insights from high volume, variety and velocity data. We split the overall process of generating data-driven insights into five stages, as shown in Fig. 1 [6].

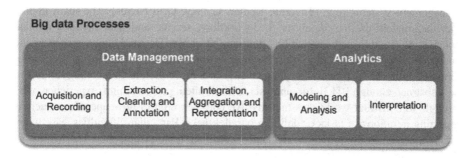

Fig. 1. Big data analytics process

In this paper, we focus on the analytics component of this big data process using statistical learning techniques. The analytics process includes modelling, analysis and interpretation of results. The theory of statistical learning was first introduced in the late 60's as the problem of function estimation from a set of data [7]. However, it took a few decades for the theories to find their way into applied science. Statistical Learning can also be described as an exploratory area of science identified by a cycle of model selection, customization and testing as scientists often do not know the exact objective or expected outcome beforehand. These days, statistical learning plays a key role in analysing data in the areas of finance, engineering, biology and science. The science of learning now is an inseparable part of artificial intelligence, data mining and statistics, and hence contributes to various disciplines [8].

Machine learning which has arisen as a field of artificial intelligence has a substantial overlap with statistical learning. Both approaches focus on supervised and unsupervised learning; Machine learning has a greater emphasis on large scale applications and predication accuracy while statistical learning emphasizes analytics models, their interpretability, precision and uncertainty. The distinction between the two has become blurred, and there is a great deal of "cross-fertilisation" [8, 9].

Some examples of statistical learning problems are:

- Forecasting the price of a company's stock in 2 years using a number of indicators such as company fundamentals and macro-economic data
- Identifying the hand-written number of ZIP codes for processing and distributing mails
- Using the infrared absorption spectrum in a person's blood to estimate the glucose level of a diabetic patients
- Using demographic and clinical data to determine the risk factor for a certain disease

As stated in the introduction, supporting multiple statistical learning models within an organisation will raise serious software and data management issues in dealing with heterogeneous platforms and data formats. In this paper, we take a higher-level view, looking at high level design i.e. *an architecture* of solutions instead of specific solutions. Multiple researchers [10–12] have advocated tackling the problem at the architectural level to provide support for a high level of automation in order to increase the efficiency and productivity of analysts. For instance, the ADAGE framework [11] leverages the capabilities of service-oriented architectures and scientific workflow management systems (e.g., Taverna, Kepler, Galaxy, Grid Nexus). The main idea is that the models used by analysts (i.e. workflow, service and data models) contain concise information and instructions that can be viewed as an accurate record of the analytics process that become a useful artefact for provenance tracking and ensuring reproducibility of such analytics processes. However, such models have proved not to be "abstract enough" [13] for analysts with limited programming experience who struggle to construct workflows that involve complex control-flow operations such as repeating iterations of tasks or defining parallel tasks. Particular difficulties arise when these analysts are required to manually establish relationships between workflow decisions, service components and data elements.

The solution advocated in this paper is to apply semantic technologies to tackle the problem of representing knowledge associated with analytics. Similar to our approach, other researchers have used semantic technologies and web ontology framework (OWL) to model knowledge in analytics space. Miller et al. [14] developed an analytics ontology to formally describe modelling techniques, models and results. Panov et al. [15] defines the OntoDM ontology, based on a recent proposal for a general frame-work for data mining and it is also aligned with the ontology of biomedical investigations. Others have focused their ontology design on algorithm selection and metalearning [16, 17].

The problem is that the proposed ontologies are always tied to a specific domain. In previous work, we have attempted to define a more generic ontology to capture the concepts related to statistical learning models [18]. The main contribution of this paper is to extend this ontology to address the representation of *variables of interest* and their inter-relationships with each other and domain objects.

3   Proposed Approach

3.1   Ontology Development Process

In this section, the ontology development process which has been used to design the Statistical Learning Ontology (SLO) is presented. The development process presented in this paper borrows the relevant principles from the various methodologies, especially the pattern-based ontology development methodologies such as the NeOn methodology [19, 20]. The ontology development process consists of identifying motivating scenarios, building the ontology and evaluating the ontology. As depicted in Fig. 2, the ontology building phase is divided into Generic Design and Domain-Specific Design activities.

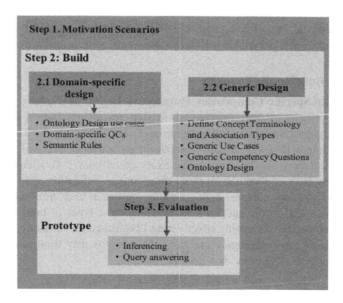

Fig. 2. Ontology engineering process

Domain specific use cases and competency questions are identified and generalized into sets of generic use cases and competency questions. The ontology building process is designed to capture knowledge sufficient to satisfy both generic and domain specific competency questions. At the evaluation phase inferencing and query answering is conducted to measure satisfiability of competency questions.

3.2   Step 1: Motivating Scenarios

As data analytics solutions are becoming a desired and integral part of many organizations we observe a need to efficiently organise different knowledge spheres related to analytics. One organization may have multitude of data sets, domain knowledge, existing analytic models and a team of data analysts who may have different opinions on how to conduct a particular analysis task, which variables to choose and how to derive those variables from raw data. Identifying data, analytics models, algorithms etc. that best fit a particular task is a daunting task for analytics solution engineers. Below is a list of applications where analytics ontology is intended to be used:

- Establish a common terminology for the organisation to represent entities and relationships in the data analytic domain. Relate to existing taxonomies, ontologies and data standards if they exist.
- Determine established (even if contradictory) complex analytic knowledge in a particular domain. The knowledge may include known relationships between different variables, how a variable is linked to a particular model, what are the relationships between variables and data sets, what are the relationships between variables and analytic models
- Integrate existing knowledge about data sets, domain, analytic models and algorithms etc. into an organisation-wide ontology to provide unified access to analytic knowledge
- Use the analytic knowledge expressed through ontology to assist data analytic process stages such as variable selection, data source selection, and model selection

3.3   Step 2: Building the Ontology

3.3.1   Domain-Specific Competency Questions – Digital Marketing

Due to the popularity of search engine advertising, researchers have a long-standing interest in studying what drives search engine advertising effectiveness. The use case in this domain intends to assess *Brand engagement on social media: will firms' social media efforts influence search engine advertising effectiveness?* To address this question researchers have conceptualised three dimensions – affiliation, conversation and responsiveness – that capture both consumer-initiated and firm-initiated brand engagement on social media and have studied how these types of brand engagement influence search engine advertising effectiveness [21]. Based on this study we identified a set of use cases and competency questions that analysts may want to ask from the ontology we plan to design, as shown in the Table 1.

Table 1. Competency questions in digital marketing domain

Use case	Competency question	Answer
A variable quantifies or represents a property	What are the variables that represent brand engagement?	Affiliation, Conversation and Responsiveness
A causal or influence relation needs to be established between two properties or variables	What variables influence search engine advertising effectiveness?	Affiliation, Search engine score
The value for a variable need to be calculated, e.g. Affiliation	What is the measure for the variable Affiliation	page_fan_count_daily (extracted from Facebook)
Identify measures that are used to calculate another measure, e.g. "page_fan_count_daily"	What are the measures used to calculate "page_fan_count_daily"?	page_fan(t) – page_fan (t-1) (extracted from Facebook)

3.3.2 Domain-Specific Competency Questions – Commodity Pricing

This case study was inspired by a Hackathon organised at University of New South Wales in partnership with ANZ Bank in Australia. The motivation is that the future success of agribusiness will be reliant on informed decisions about capacity, investment and other driving factors. Many of the banks' customers are interested in questions like "which countries and consumers will buy our products? what prices and economic value is likely to be generated from this? what primary or processed food products should Australia seek to produce in future?". The idea of the competition was to use public and private data on this sector – macro-economic indicators, production volumes, weather patterns, prices, etc. to investigate what will drive this industry going forward [22]. Table 2 represent the set of competency questions we identified for this case study.

Table 2. Competency questions in commodity pricing domain

Use case	Competency question	Answer
Variables that are correlated to a given variable	What are the variables correlated with GDP growth in China?	Urbanisation Growth, Interest rate, Exchange rate
The models that establish a relationship between two variables	What are the models and their inputs and outputs, which can predict GDP in China?	A list of Statistical Learning models including input/outputs for each model, e.g. "multiple regression" as model uses "urbanization rate in China" and "interest rate in China" to predict "GDP in China"
Calculate the value/measure of a variable or property using models	How does a 5% increase in urbanization in China can influence GDP in China?	Plug in 5% as input to the model to calculate the measure for GDP in China
Identify measures that are used to calculate another measure, e.g. "GDP_Growth_China"	What is the measure used to calculate "GDP_Growth_China"	Consumption_China – Invesment_China + Government_Expense_China

3.3.3   Generic Concepts and Association Types

Based on the common terminologies identified through domain specific competency questions we defined a set of key concepts to eliminate ambiguity across different communities. These concepts are listed in the Table 3.

Table 3. Definitions of generic concepts

Generic concept	Definition
Thing	Abstraction of any entity which may exist in an ontology
Property	Any measurable characteristic of a thing
Variable	Metrics which are used to represent a certain property of a thing
Measure	Actual value that a variable takes
Model	Statistical learning model which is used to establish relationships

The next step in ontology engineering is the definition of association types that link different concepts together. Main association types relevant to our ontology are how the variables are linked with each other. We identified two such types: 1. Causal or inferable relationship and 2. Hypothesized relationship that represents a hypothetical link between variables that needs to be tested. These two link types can be originated in a number of ways. To represent how a relationship was originated, for provenance and understanding, we define a set of concepts called link origins, which are shown in Table 4.

Table 4. Set of origins for variable association

Link origin	Definition
Opinion	The link between two variables is the opinion of an expert or an individual, and thus established
Model	The link between two variables were established via a model
Reference	The link between two variables were established in literature

3.3.4   Generic Competency Questions

Based on the domain specific competency questions and the terminology we defined in the Sect. 3.3.3, we designed a set of generic (domain independent) competency questions as shown in the Table 5. The generic competency questions aggregate the outputs of domain-specific competency question, and list them using the generic concepts defined in Tables 3, and 4.

Table 5. Generic competency questions

Use case	Competency question	Answer
The variables linked to a given variable through model	What are the variables linked through a model with a certain variable v_1?	v_2 & v_3 influence v_1 through model m_1 v_4 influence v_1 through model m_2
The model that establishes a link between two properties or variables	What are the statistical learning models which create a link between a certain variable and other variables?	List of models with inputs for each model, for instance: model m_3 links v_1 to v_4 (causal link)
Measures that calculate the value of a variable	What is the measure that records the value of a particular variable?	A raw measure or a processed measure
Identify measures which are used to calculate another measure	What are the measure used to calculated a certain measure	A list of raw measures or processed measures that are used to calculate a desired measure
Identify the link origins that exist between two variables	What is the origin of the link between v_1 and v_2?	Opinion, model or reference
Identify link types between two variables	What are the possibilities for link types between v_1 and v_2?	Causal or Hypothetical link types

3.3.5   The Statistical Learning Ontology (SLO)

Based on the generic competency questions, we designed the statistical learning ontology. The components of the ontology are shown in Fig. 3. Main classes of the ontology are Variable, Measure, Model and LinkedVariables. LinkedVariables class captures a link between any two variables. LinkType and LinkOrigin are two classes we define to represent details of different links between variables. We extended those two classes via sub-class hierarchy and created an extendable catalogue of link types (Causal, Hypothesized) and origins (ViaModel, ViaLiterature, and FromModels).

One strength of this ontology is how it can be linked with any domain specific ontology via an operationalised relationship. We can take any concept (thing or property) defined in any third-party ontology and link it to a variable used for statistical learning. In this way, the context of the variable available in domain specific ontology is readily available to the SLO. Note that the prefix ano in the ontology refers to the SLO.

We adapt the RDF-Cube vocabulary (presented by the prefix "qb") to represent datasets and link ano:Measure concept with RDF-Cube qb:MeasureProperty. This way our ontology can be easily integrated into existing linked datasets defined in RDF-Cube structure.

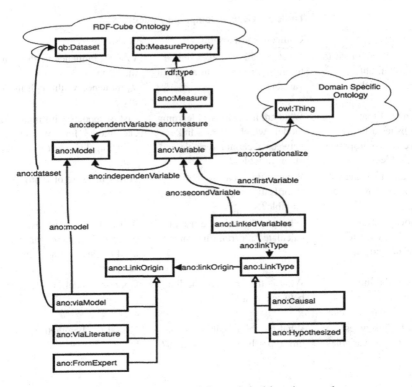

Fig. 3. Main components of the statistical learning ontology

4 Prototype Application for SLO

4.1 Introduction

To evaluate the SLO and demonstrate its capabilities we developed a web based application prototype, following the architecture defined in Fig. 4.

The core of the architecture is the SLO. Different knowledge packs can be created integrating SLO and existing domain ontologies to capture analytic knowledge related to each domain. An instance repository is created by acquiring data sets in RDF-Cube format and linking them to different measures defined in the knowledge pack.

SLO, knowledge packs and the instance data were created as RDF triples following semantic web standards. For the creation of the RDF triples, integrating existing domain ontologies we used the popular ontology editing tool- Protégé[1]. Created triples are stored in a semantic repository. For the prototype, we used MarkLogic triple store[2].

---

[1] https://protege.stanford.edu.

[2] https://www.marklogic.com/product/marklogic-database-overview/database-features/semantics/.

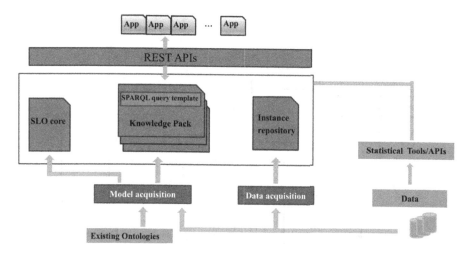

Fig. 4. Architecture of SLO

We created a REST API baked by semantic models to query and retrieve knowledge from the core ontology, knowledge packs and instance data. SPARQL templates were defined to retrieve information related to each use case which is accessed through REST APIs.

Finally, we developed a web application that enables users to interact with SLO via web service APIs. The user is presented with a set of knowledge packs when the application realizing the use cases is initiated. Once the user selects an appropriate knowledge pack for their domain of interest, it will be accessed to answer user queries.

4.2 Realising the Use Cases

Assisting analysts to conduct statistical analysis in the Digital Marketing and Commodity Pricing domains are the two areas we evaluate through this prototype. List of use cases for each domain are presented in the first column of Table 1 and Table 2 in Sects. 3.3.1 and 3.3.2 respectively.

We used SLO to populate knowledge packs in Digital Marketing and Commodity Pricing domains. Then the web application accesses those knowledge packs via semantic queries to retrieve and present information to users.

4.3 Knowledge Pack for Digital Marketing

Figure 5 demonstrates a snapshot of the knowledge pack created for the Digital Marketing case study.

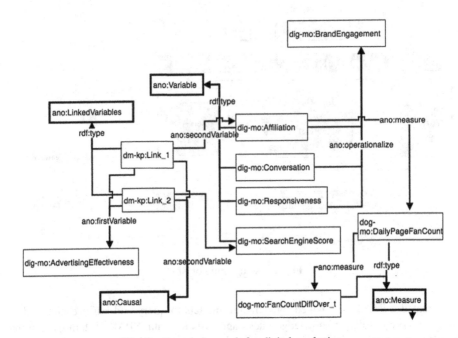

Fig. 5. Knowledge pack for digital marketing

As there are no publicly available ontologies that define concepts related to digital marketing, we created a simple Digital Marketing Ontology (dig-mo) for the case study. The "dig-mo" prefix is used to represent concepts (Properties and Things) defined in the domain ontology. "dm-kp" is the prefix used to represent concepts defined specific to the digital marketing knowledge pack. One example is the instance created from ano:LinkedVariable class- dm-kp:Link_1 that defines influence as a relationship (Causal) of Affiliation to the SearchEngineScore variable.

4.4 Answering Selected Competency Questions via Semantic Queries

In this section, we demonstrate the set of SPARQL queries we can use to answer competency questions defined in Table 1 for digital marketing. We have implemented a set of query templates in the backend so users can select any variable/concept or measure from the front-end application that relate to competency questions.

The queries implemented support the following questions:

1. What are the variables that represent brand engagement?

PREFIX dig-mo: http://www.adage.unsw.edu.au/digital-marketing
PREFIX ano: http://www.adage.unsw.edu.au/analytics
SELECT ?variable
        WHERE { ?variable ano:operationalize dig-mo:BrandEngagement
                }

2. What variables influence (cause) search engine advertising effectiveness?

```
PREFIX ano: http://www.adage.unsw.edu.au/analytics
PREFIX dig-mo: http://www.adage.unsw.edu.au/digital-marketing

SELECT ?variable
    WHERE {
                ? linkedVariable ano:firstVariable dig-mo:AdvertisingEffectiveness.
                ? linkedVariable ano:secondVariable ?variable.
                ?linkedVariable ano:linkType ano:Causal.
        }
```

3. What is the measure for the variable Affiliation

```
PREFIX ano: http://www.adage.unsw.edu.au/analytics
PREFIX dig-mo: http://www.adage.unsw.edu.au/digital-marketing

SELECT ?measure
    WHERE {
                Dig-mo:Affiliation ano:measure ?measure.
        }
```

4. What are the measures used to calculate "page_fan_count_daily"?

```
PREFIX ano: http://www.adage.unsw.edu.au/analytics
PREFIX dig-mo: http://www.adage.unsw.edu.au/digital-marketing

SELECT ?measure
    WHERE {
                Dig-mo:DailyPageFanCount ano:measure ?measure.
        }
```

4.5 Visualizing Variables

Our application can visualize all the variables and how they are interlinked for a user-selected knowledge pack. The following SPARQL query is used to fetch all the data necessary for visualization.

```
PREFIX ano: <http://www.adage.unsw.edu.au/analytics>
SELECT ?variable1 ?variable2 ?linkOrigin ?linkType

    WHERE {
              ? linkedVariable ano:firstVariable ?variable1.
              ? linkedVariable ano:secondVariable ?variable2.
              ?linkedVariable ano:linkOrigin ?linkOrigin.
              ?linkedVariable ano:linkType ?linkType
        }
```

Figure 6 shows the visualization generated in the application from the data returned by the query above. Different node types represent variables, concepts and context. When the user hovers over an edge, the link type and source (origin) of that link are displayed in the box below. In this example, the link on how the Conversation variable influences the Conversion Rate variable is shown. The link origin is indicated as model in addition to the reference from which this link is established.

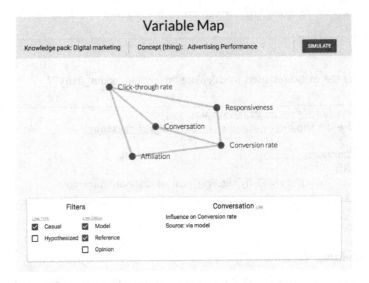

Fig. 6. Visualisation of variables

The Simulate button allow users to rearrange the graph, Filters are used to restrict the visualization by link types and origins. For example, user can use a filter to show only the causal links.

5   Conclusions and Future Work

This paper proposes a model-driven architecture that enhances knowledge sharing by data analysts and domain experts in an exploratory way in spite of heterogeneity and complexity. The focus of this paper is the design of an ontology which encompasses three main features: (1) it captures the concepts in statistical learning algorithms and explicitly defines the relationships between variables, properties, measures and models (2) it is an open semantic solution which will leverage exiting domain-specific ontologies and also allow different types of statistical learning techniques to be added (3) it provides a generic research environment that can be adapted to different scientific domains by customizing the application. The semantic web infrastructure enables resource sharing, integration and self-improvement through collaborative scientific research.

Acknowledgements. We are grateful to Capsifi and Ignition Wealth, especially Terry Roach, Mark Fordree and Mike Giles for sponsoring the research which led to this paper. We are also grateful to Adnene Guabtni and Chedia Dhaoui for helping with digital marketing case study. We thank Gino Conte on the visualization development of the prototype application.

References

1. The Computational Social Science Society of the Americas. https://computationalsocial-science.org
2. Gilbert, N. (ed.): Computational Social Science, vol. 21. Sage, Thousand Oaks (2010)
3. Dhar, V.: Data science and prediction. Commun. ACM **56**(12), 64–73 (2013)
4. Schlegal, K., Linden, A.: Predicts 2017: Analytics Strategy and Technology. Gartner, Stamford (2016)
5. Nural, M.V., Cotterell, M.E., Miller, J.A.: Using semantics in predictive big data analytics. In: 2015 IEEE International Congress on Big Data (BigData Congress), pp. 254–261. IEEE, June 2015
6. Labrinidis, A., Jagadish, H.V.: Challenges and opportunities with big data. Proc. VLDB Endow. **5**(12), 2032–2033 (2012)
7. Vapnik, V.N.: An overview of statistical learning theory. IEEE Trans. Neural Networks **10** (5), 988–999 (1999)
8. Migon, H.S., Gamerman, D., Louzada, F.: Statistical Inference: An Integrated Approach. CRC Press, Boca Raton (2014)
9. James, G., Witten, D., Hastie, T., Tibshirani, R.: An Introduction to Statistical Learning, vol. 112. Springer, New York (2013)
10. Harper, K.E., Dagnino, A.: Agile software architecture in advanced data analytics. In: 2014 IEEE/IFIP Conference on Software Architecture (WICSA), pp. 243–246. IEEE, April 2014
11. Yao, L., Rabhi, F.A.: Building architectures for data-intensive science using the ADAGE framework. Concurr. Comput. Pract. Exp. **27**(5), 1188–1206 (2015)
12. Behnaz, A., Rabhi, F., Peat, M.: A software architecture for enabling statistical learning on big data. In: Rojas, I., Pomares, H., Valenzuela, O. (eds.) ITISE 2016. CS, pp. 343–357. Springer, Cham (2017). https://doi.org/10.1007/978-3-319-55789-2_24

13. Withers, D., Kawas, E., McCarthy, L., Vandervalk, B., Wilkinson, M.: Semantically-guided workflow construction in Taverna: the SADI and BioMoby plug-ins. In: Margaria, T., Steffen, B. (eds.) ISoLA 2010. LNCS, vol. 6415, pp. 301–312. Springer, Heidelberg (2010). https://doi.org/10.1007/978-3-642-16558-0_26

14. Miller, J.A., Han, J., Hybinette, M.: Using domain specific language for modeling and simulation: scalation as a case study. In: Proceedings of the Winter Simulation Conference, pp. 741–752, December 2010

15. Panov, P., Džeroski, S., Soldatova, L.: OntoDM: an ontology of data mining. In: IEEE International Conference on Data Mining Workshops, ICDMW 2008, pp. 752–760. IEEE, December 2008

16. Lin, M.S., Zhang, H., Yu, Z.G.: An ontology for supporting data mining process. In: IMACS Multiconference on Computational Engineering in Systems Applications, vol. 2, pp. 2074–2077. IEEE, October 2006

17. Espinosa, R., García-Saiz, D., Zorrilla, M.E., Zubcoff, J.J., Mazón, J.N.: Development of a knowledge base for enabling non-expert users to apply data mining algorithms. In: SIMPDA, pp. 46–61, August 2013

18. Behnaz, A., Natarajan, A., Rabhi, Fethi A., Peat, M.: A semantic-based analytics architecture and its application to commodity pricing. In: Feuerriegel, S., Neumann, D. (eds.) FinanceCom 2016. LNBIP, vol. 276, pp. 17–31. Springer, Cham (2017). https://doi.org/10.1007/978-3-319-52764-2_2

19. Shah, T.M.: Designing and conceptualising ontology patterns for modelling cross-domain health information. Ph.D. thesis, University of New South Wales (2016)

20. Suárez-Figueroa, M.C.: NeOn methodology for building ontology networks: specification, scheduling and reuse. Doctoral thesis, Artificial Intelligence, Universidad Politécnica De Madrid (2010)

21. Yang, S., Lin, S., Carlson, J.R., Ross Jr., W.T.: Brand engagement on social media: will firms' social media efforts influence search engine advertising effectiveness? J. Mark. Manage. 32(5–6), 526–557 (2016)

22. Info Package for UNSW Data Science Hackathon. http://www.cse.unsw.edu.au/~fethir/HackathonInfo/HackathonStudentPack_v7.pdf

Author Index

Printed in the United States
By Bookmasters